Handbook
of
Model
Rocketry

SEVENTH EDITION

Handbook of Model Rocketry

**G. HARRY STINE
AND
BILL STINE**

WILEY

John Wiley & Sons, Inc.

Published by John Wiley & Sons, Inc., Hoboken, New Jersey
Published simultaneously in Canada

Originally published by Follett Publishing Company, Chicago
Previously published by Arco Publishing, Inc., New York, and Prentice Hall Press, New York

For general information about our other products and services, please contact our Customer Care Department within the United States at (800) 762-2974, outside the United States at (317) 572-3993 or fax (317) 572-4002.

Wiley also publishes its books in a variety of electronic formats. Some content that appears in print may not be available in electronic books. For more information about Wiley products, visit our web site at www.wiley.com.

Library of Congress Cataloging-in-Publication Data:
Stine, G. Harry (George Harry), date.
  Handbook of model rocketry / G. Harry Stine and Bill Stine.— 7th ed.
      p.   cm.
  Includes bibliographical references and index.
  ISBN 0-471-47242-5 (pbk.: alk. paper)
  1. Rockets (Aeronautics)—Models. I. Stine, Bill. II. Title.
  TL844.S77 2004
  629.46'52'0228—dc22
                                                    2004002230

Printed in the United States of America

10 9 8 7 6 5 4 3 2 1

CONTENTS

PREFACE TO THE SEVENTH EDITION

Few people have the opportunity to carry forward a legacy. I am one of the fortunates. Model rocketry is both my hobby and my livelihood. On my eighteenth birthday, my father wrote in my personal copy of the fourth edition of the *Handbook of Model Rocketry* the following: *Bill, you don't have to carry forward what I've started here, but you must do something meaningful with your life—I know you will. . . . Love, Dad.* And since that day, I've pursued my love of model rockets as a profession.

Model rocketry was to kids in the sixties and seventies what the Internet and computer games are to today's generation. Harry was a great visionary and saw the social need in 1957 to create model rocketry. And from this creation we have witnessed generations of young people turned on to science, math, and aerospace careers. There are now more astronauts, engineers, and aerospace industry leaders who flew model rockets as youngsters than ever before.

The personal computer and the Internet have forever changed this hobby. When this book was first written in 1963, a young person who showed an interest in anything would go to school or the local library and search out any books on the subject. The first several editions were literally devoured by young aspiring rocketeers hungry for information. Today we have the Internet, whose numerous forums and Web sites disseminate information quickly all over the world. I often see postings such as "Has anyone else had difficulty getting such and such to work?" And there's always several answers posted within hours from fellow rocketeers!

I hope that this revised edition will remain such a highly regarded reference. The easy access to software that calculates CP and CG and altitude does not replace the *theory* behind those calculations. I've purposely left in all the pencil-and-paper calculations so that readers can understand *how* a software program is doing them.

Model rocketry is unique in that you can design something with a computer, then actually build it and test it very easily and inexpensively. It is the synthesis of the old method with the new. Socratic learning at its best—design it, try it out, learn from your mistakes, and try it again!

I have taught several thousand schoolteachers and youth group leaders how to use model rocketry as a teaching tool. I believe it is a learning process in disguise. It excites and engages young people and makes it easy to teach basic math and science while having fun. I've never failed to see a young person's face light up with joy as he or she launched his or her rocket and dreams skyward.

I am considered an expert on model rocketry. But I've learned that "knowledge is the enemy of learning." In other words, if you think you already know it all, your mind becomes closed to learning anything new. Read this book, learn something, and if at all possible, go ahead and "pay it forward" (pass your knowledge on).

BILL STINE
Cave Creek, Arizona
2004

PREFACE TO THE FIRST EDITION

Every author has a book or story that he or she has always wanted to write. In my case, this is it. I labor under the delusion that it might be more important than all the fiction I've done and of broader consequence than much of my nonfiction.

I hope that it may save the hands, eyes, and lives of countless youngsters who might never have learned about model rocketry without it. I also hope it may set many young people on their course toward becoming astronauts, engineers, technicians, and other kinds of scientists. Finally, I hope it may serve as a guidepost to many people, young and old, who are interested in rockets.

Model rocketry is my hobby. I have given much time and effort to it and I have gained rewards far more valuable than mere money from it. The basic motive for my involvement in model rocketry stems from my youth, when many scientists, engineers, teachers, and other adults freely gave me advice, guidance, help, and the means to do things.

Once I asked one of these people what I could do to repay him. "Do the same for others when you grow up," he told me. In the Space Age hobby of model rocketry, I have found a way to do this.

G. HARRY STINE
New Canaan, Connecticut
1963

THIS IS MODEL ROCKETRY

Model rocketry has been called miniature astronautics, a technology in miniature, a hobby, a sport, a technological recreation, and an educational tool—and it is *all* of these things. It is a safe, enjoyable, and highly respected pastime that now boasts the enthusiastic participation of millions of people, young and old, in countries such as Australia, Bulgaria, Canada, China, the Czech and Slovak Republics, France, Germany, Japan, Latvia, the Netherlands, Poland, Romania, Russia, Slovenia, Spain, Switzerland, the Ukraine, the United Kingdom, and the United States.

Model rocketry started in the United States in 1957. Its beginnings were carefully and thoroughly documented. Models, correspondence, drawings, publications, and other artifacts and documents concerning the precise details of the early years have been donated to the National Air and Space Museum of the Smithsonian Institution in Washington, D.C., where this extensive collection can be properly preserved for posterity. The sense of history, of doing something new and unique in the world, ran high among those of us who participated in those early years. As a result, an enormous collection of significant things was saved.

Model rocketry resulted from a timely combination of model aviation, the ancient art of pyrotechnics, and modern rocket technology. Although all of these elements had existed separately for over a decade before 1957, it fell to two people to combine them successfully into a Space Age hobby.

The first model rockets were built and flown by the late Orville H. Carlisle, the owner of a shoe store in Norfolk, Nebraska. Carlisle was an outstanding amateur pyrotechnician. His brother, Robert, was a model airplane enthusiast. With Robert's help, Orville designed the first model rockets in 1954. Early in 1957, he wrote to me, and I added the elements of professional rocket technology. Thus, today's model rocketry was born.

What is model rocketry? What makes model rockets so safe, so inexpensive, so easy to build, and so much fun to fly that millions of people have

FIGURE 1-1 Model rocketry has been a national and international sport and hobby since 1957. Here a model rocket lifts off from the Mall in Washington, D.C., with the nation's Capitol in the background. The occasion was the annual Mall demonstration of the National Association of Rocketry (NAR) sponsored by the National Air and Space Museum of the Smithsonian Institution.

successfully launched them since the first kits and motors appeared in April 1958? Why has model rocketry brought the Space Age to Main Street, directly involving more people in rocket technology than have ever watched a real space rocket launch from Cape Canaveral?

A model rocket is an aerospace model, a miniature version of a real space vehicle, with *all* of the following characteristics:

1. It is made of paper, balsa wood, plastic, cardboard, and other non-metallic materials used as structural parts.

FIGURE 1-2   Model rocketry was started by the late G. Harry Stine (left) and the late Orville H. Carlisle (right), founders of the National Association of Rocketry, shown here comparing model rockets at the Ninth Annual National Model Rocket Championships in 1967.

2. It weighs less than 3.3 pounds (1,500 grams) and uses less than 4.4 ounces (125 grams) of rocket propellant.
3. It uses professionally designed, factory-made solid-propellant rocket motors produced to strict national and international standards for reliability. In model rocketry, we don't make our own rocket motors.
4. Its solid-propellant rocket motor is ignited electrically from a distance of 15 feet or more using a battery and an electric launch controller with built-in safety features that allow the motor to be ignited on a countdown only when conditions are safe for flight.
5. It contains a recovery device (or devices) to lower it gently and safely back to the ground; thus, it can be flown again and again by installing a new solid-propellant rocket motor and repacking the recovery device.

A model rocket is that simple, yet it changed the nature of nonprofessional rocketry. Before 1958, nonprofessional rocketry was so dangerous that it was banned by law in most states. The American Rocket Society (now the American Institute of Aeronautics and Astronautics) estimated that one out of seven nonprofessional rocketeers would be injured or killed in their avocation. Today, model rocketry is so ubiquitous, so available, and so safe that you can launch almost anywhere with safety, provided you follow the simple rules of

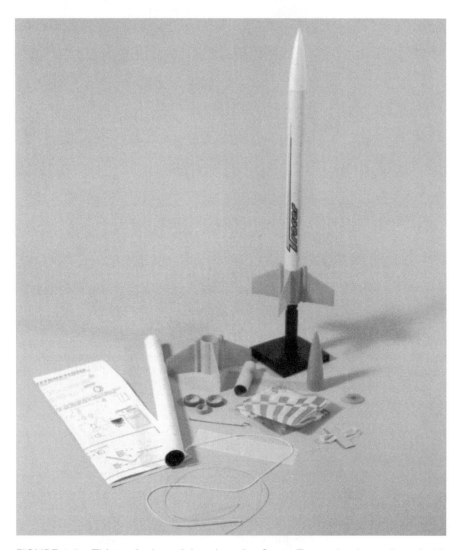

FIGURE 1-3   This typical model rocket, the Quest Tracer, is shown here in kit and assembled forms. It uses a plastic nose and tail fin unit, paper body, and parachute recovery.

the National Association of Rocketry (NAR) Model Rocket Safety Code. Because of its outstanding safety record, model rocketry enjoys the support of such government agencies as the Federal Aviation Administration (FAA), the Consumer Product Safety Commission, the National Aeronautics and Space Administration (NASA), and the armed services. It has won the support of the prestigious and safety conscious National Fire Protection Association (NFPA). Model rocketry is used by the Boy Scouts of America, the 4-H Clubs of America, the YMCA, the Civil Air Patrol, the Young Astronauts, and the various

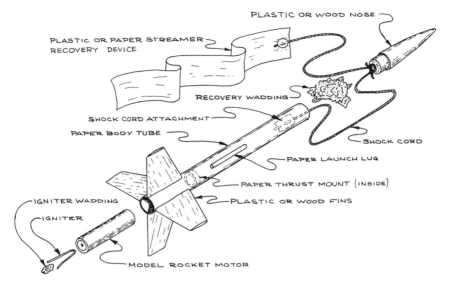

FIGURE 1-4   A sketch of a typical model rocket showing its various parts. There are many variations in the size and shape of the parts, but all model rockets are basically similar.

space camps. Model rocketry is used as a learning tool in more than 20,000 elementary and secondary schools. Even the American Institute of Aeronautics and Astronautics reversed its early stand against nonprofessional rocketry to endorse model rocketry.

This confidence is deserved. As of 2003, nearly 600 million model rockets have been flown safely and successfully in the United States alone. The hobby is far safer than swimming, boating, baseball, football, and cycling. Since 1964, members of the National Association of Rocketry (NAR) have been covered by a major liability insurance policy underwritten by a large domestic insurance firm.

As of this updated writing in 2003, all fifty states have adopted permissive state laws based on the National Fire Protection Association's NFPA 1122 "Code for Model Rocketry." This permissive national standard requires extensive product testing and quality control of model rocket motors by the manufacturers; certification by the NAR; and no permits for sale, purchase, possession, or use by merchants or the public. The only exceptions are that you must be 14 to purchase ¼A–D motors in California or ¼A–C motors in New Jersey. You must be 18 to purchase larger motors in California and New Jersey. It is always a good idea to check with your local fire authority before launching a model rocket within your city's limits.

Naturally, it's possible to get hurt in model rocketry if you're stupid, don't follow the NAR Model Rocket Safety Code, and don't read and follow instructions. It's possible to get hurt with *any* sort of technology, new or old. Some people even manage to get hurt while collecting stamps.

Model rocketry's excellent safety record is due primarily to the premanu-factured model rocket motor, the lightweight construction of the model rocket airframe, and the voluntary limitations that model rocketeers have placed on their hobby.

It's not obvious at first that you can learn anything about rockets if you can't experiment with the rocket motor itself. But the science and technology of astronautics are more than a matter of rocket motors and propulsion. The business of making rocket propellants as well as designing and making a rocket motor of *any* type or size is a very complicated, expensive, dangerous, and delicate affair that must *not* be attempted by anyone with less than many thousands of dollars for the proper equipment, an advanced college degree in chemistry, several years' experience in handling explosive materials, several acres of land providing a safe place to work, and a very large life insurance policy. (Very few professional model rocket motor technicians have been killed while making model rocket motors under the most carefully controlled conditions with all of the safety equipment available. The model rocket motor manufacturing companies therefore take grave risks and assume the awful hazards of rocket motor manufacture so that model rocketeers can enjoy the hobby safely.)

Model rocketry is very similar to model aviation. If you want to build a flying model airplane, it would be very expensive and time-consuming if you had to build the *entire* model, including the little gasoline engine. You have more fun and more flying time if you build only the model airplane's airframe and purchase the motor ready-to-run. Then you know you have a motor that will work and—if you're a careful modeler and flyer—an airplane that will usually get high enough to crash.

The same holds true for model rocketry. By properly using a factory-made model rocket motor and building a model rocket airframe around it, you can have an excellent chance of flying it successfully and safely. Furthermore, you will eventually learn about rocket thrust, duration, total impulse, specific impulse, grain configuration, thrust-time performance, and other rocket motor technology. Since a model rocket is a free body in space even when it's in the air, its actions in flight are quite different from those on the ground on wheels or skids. Subtract the known effects of the surrounding air and you will be able to understand how a rocket behaves in the vacuum of space. Flight performance can tell you a great deal about how things move in the universe. And even the fact that your model rocket always flies within the earth's atmo-sphere will introduce you to the fascinating mysteries of aerodynamics, weight and balance, stability, and drag. In finding out more about why your model rockets fly as they do (and sometimes why they don't fly as predicted), you can begin to delve into optics, structures, dynamics, electronics, meteorology, materials science, and many other modern technologies including the mathe-matical tools that make them useful to humankind. Model rocket design and flight performance are also excellent areas where computer analysis can be brought into play. Many good computer programs now exist that permit a model rocketeer to use a personal computer or personal digital assistant to determine model rocket design and flight factors. You'll learn more about these in chapters 8 and 9.

A model rocket is wonderfully simple to build. You can use common hobby tools of the sort used to build model airplanes, model cars, and model boats. Even the materials employed in these other modeling hobbies are used in model rocketry. Many people believe that a rocket has to be made out of steel or other metals, but this isn't so. Why make a rocket out of metal when you can make one cheaply out of paper, plastic, and wood? Why spend a lot of money to buy a metal-welding outfit when a tube of white glue or epoxy will do the same job?

Besides ease of construction, there are other reasons for using nonmetallic materials. High strength and light weight have always been prime design goals for flying devices of any type. Paper, plastic, and balsa wood meet these requirements well. Even the new composite materials are now being used in model rocketry (paper, of course, being an ancient composite material). All of these materials are surprisingly strong. In fact, balsa wood has a higher ratio of strength to weight than carbon steel.

In addition, nonmetallic materials used as structural parts—nose, body, or fins—are much safer should something go wrong during the flight of a model rocket. A model rocket literally disintegrates when it hits something because its airframe absorbs the energy of impact by destroying itself. This is the same principle used in modern automobiles where "crush zones" absorb the energy of a crash by deforming and collapsing. Model rockets have been deliberately launched directly into sheets of window glass; these experiments completely destroyed the models but didn't even scratch the glass. (The tests were done by professionals to find out what would happen; please don't try this in your living room!)

All model rockets have recovery devices. You might ask, "Why bother about a recovery device?" Answer: because you want your model rocket to come down and land as gently as possible in a condition to fly again, just like the NASA space shuttle *Orbiter*. After you've spent a lot of money and taken hours of time to build a model rocket, you'll want to get more than a single flight out of it. Recovery devices are easy to install. And they work with exceedingly high reliability. Furthermore, the requirement for a recovery device helps make model rocketry a safe activity. Even a 2-ounce model rocket plummeting back to the ground from an altitude of 500 feet can hurt if it hits you. A recovery device to retard the speed of its descent eliminates this potential hazard. The mandatory use of recovery devices is one of the reasons model rocketry has been and continues to be such a safe activity.

But why build model rockets so small? Why not build them big so that they're impressive, go several miles high, or carry lots of payload like the big ones at Cape Canaveral? There are several reasons. Model rockets are small for the sake of cost, simplicity, and safety. The price of a model rocket goes up even faster than the rocket itself as the size increases. You can look at any model rocket catalog for proof of this. Big models get to be very expensive. And, as size increases, they become more difficult and demanding to build. Large model

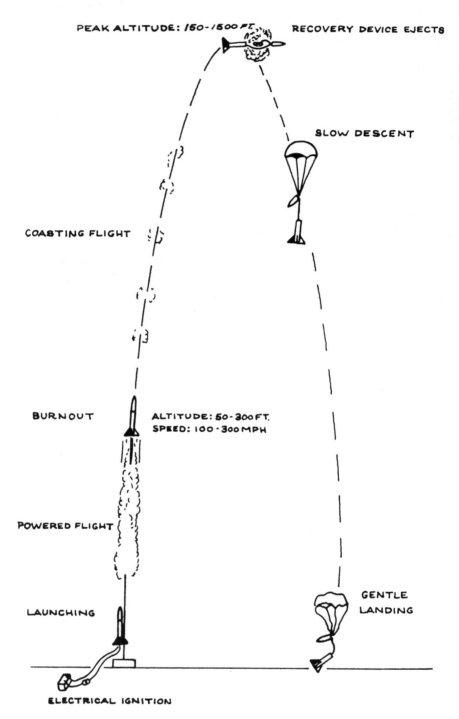

PEAK ALTITUDE: *150-1500 FT.*   RECOVERY DEVICE EJECTS

SLOW DESCENT

COASTING FLIGHT

BURNOUT   ALTITUDE: 50-300 FT.
           SPEED: 100-300 MPH

POWERED FLIGHT

LAUNCHING   GENTLE
             LANDING

ELECTRICAL IGNITION

FIGURE 1-5   The typical flight of a single-staged model rocket with parachute recovery.

rocket motors are required for large model rockets, and large model rocket motors are expensive. Some adults have become active in high-power rocketry—building very large rockets powered by very large and very expensive rocket motors. These require open land areas several square miles in extent, plus a waiver from the FAA for flight. I recall watching one large high-power rocket fail in flight on a dry lake bed in the desert of the American West; the rocketeer moaned, "There goes six months of work and $1,200!" (For that amount of money and in that time period, I can fly a *lot* of model rockets that will actually go higher!) That rocketeer might just as well go to work for NASA, the armed forces, or an aerospace company and get paid for it.

To begin, you'll work hard getting your first model rocket together and off the ground for its first successful flight. You can do it—*if you read and follow all instructions carefully!*

If at first you don't succeed, try reading the instructions.

After your first launch or two, you'll be striving for reliable flight, reliable ignition, straight flight paths, full recovery deployment, recovery in the same county, getting the model out of trees, and getting it ready for another flight. You'll progress to more difficult models, motors of higher power, multistage models, glide-recovery models, and payload-carrying models. You'll be able to try your hand at scale model rockets. Or perhaps you'll design your own model and experience the thrill of seeing it work exactly as you predicted. Maybe you'll enter contests, and, after learning all the ins and outs of model rocket competition, start to win ribbons, trophies, and prizes. And you'll make lots of new friends who share your interest in model rocketry. Some of these people will turn out to become friends for life.

As you grow more deeply involved in model rocketry, you'll find yourself studying a wide variety of subjects in order to understand what your model rockets are doing and how to improve them. You'll *never* have to worry about a science fair project if you're a student rocketeer.

When a model rocketeer talks with a professional rocketeer, they speak the same language. They can talk about specific impulse, ballistics, data reduction, static testing, spin stability, recovery parachute deployment, lift-to-drag ratios, and hundreds of other topics. And, surprisingly, they have a great deal of mutual respect for one another. Many professional rocket engineers are also model rocketeers. And now, after half a century, many young people who started as model rocketeers have become professionals in the aerospace sciences and engineering fields. I can hardly go to a NASA center or an aerospace company without meeting somebody whom I last saw years before on a model rocket range, as a teenage student taking the first steps toward the stars.

This is because model rocketry is indeed a technology in miniature and was deliberately designed that way. It recognizes that there are many interesting aspects to aerospace technology other than the propulsion system. To model rocketeers, a model rocket motor is nothing more than a prime mover, a realistic device that produces the thrust force to lift their models into the sky.

Model rocketry has become many things to many people. It can be a means to learn something about the universe. It can be a way to satisfy the

FIGURE 1-6 Model rocketry is international, having spread from the United States to around the world. Here Juliuz Jaronzyk of Poland puts his scale model of the French Diamant satellite launch vehicle on the pad with the help of his fellow rocketeers.

TABLE 1
NAR Model Rocket Safety Code (Approved February 10, 2001)

1. *Materials.* I will use only lightweight, nonmetal parts for the nose, body, and fins of my rocket.

2. *Motors.* I will use only certified, commercially made model rocket motors and will not tamper with these motors or use them for any purposes except those recommended by the manufacturer.

3. *Ignition System.* I will launch my rockets with an electrical launch system and electrical motor igniters. My launch system will have a safety interlock in series with the launch switch, and I will use a launch switch that returns to the "off" position when released.

4. *Misfires.* If my rocket does not launch when I press the button of my electrical launch system, I will remove the launcher's safety interlock or disconnect its battery and will wait 60 seconds after the last launch attempt before allowing anyone to approach the rocket.

5. *Launch Safety.* I will use a countdown before launch and will ensure that everyone is paying attention and is a safe distance of at least 15 feet away when I launch rockets with D motors or smaller and 30 feet away when I launch larger rockets. If I am uncertain about the safety or stability of an untested rocket, I will check the stability before flight and will fly it only after warning spectators and clearing them away to a safe distance.

6. *Launcher.* I will launch my rocket from a launch rod, tower, or rail that is pointed to within 30 degrees of the vertical to ensure that the rocket flies nearly straight up, and I will use a blast deflector to prevent the motor's exhaust from hitting the ground. To prevent accidental eye injury, I will place launchers so that the end of the launch rod is above eye level or will cap the end of the rod when it is not in use.

7. *Size.* My model rocket will not weigh more than 1,500 grams (53 ounces) at liftoff and will not contain more than 125 grams (4.4 ounces) of propellant or 320 newton-seconds (71.9 pound-seconds) of total impulse. If my model rocket weighs more than 1 pound (453 grams) at liftoff or has more than 113 grams (4 ounces) of propellant, I will check and comply with Federal Aviation Administration regulations before flying.

8. *Flight Safety.* I will not launch my rocket at targets, into clouds, or near airplanes, and I will not put any flammable or explosive payload in my rocket.

9. *Launch Site.* I will launch my rocket outdoors, in an open area at least as large as shown in the accompanying table, and in safe weather conditions with wind speeds no greater than 20 miles per hour. I will ensure that there is no dry grass close to the launch pad and that the launch site does not present the risk of grass fires.

Launch Site Dimensions

Installed Total Impulse (N-sec)	Equivalent Motor Type	Minimum Site Dimensions (ft.)
0.00–1.25	$\frac{1}{4}$A, $\frac{1}{2}$A	50
1.26–2.50	A	100
2.51–5.00	B	200
5.01–10.00	C	400
10.01–20.00	D	500
20.01–40.00	E	1,000
40.01–80.00	F	1,000
80.01–160.00	G	1,000
160.01–320.00	2Gs	1,500

(continues)

TABLE 1
(continued)

___

10. *Recovery System.* I will use a recovery system such as a streamer or parachute in my rocket so that it returns safely and undamaged and can be flown again, and I will use only flame-resistant or fireproof recovery system wadding in my rocket.

11. *Recovery Safety.* I will not attempt to recover my rocket from power lines, tall trees, or other dangerous places.

___

competitive spirit that says, "My model is better than your model!" It can be a way to learn and a way to teach others. For many people, it's an enjoyable recreation that combines the individualistic craftsmanship of the home work-shop and the happy socializing with others on the flying field in the sunshine and fresh air. It can be a way for young and old to get together in an activity that interests both.

Model rocketry combines modern science and technology, craftsmanship and shop practice, individual creativity and group cooperation, and the pur-suit of excellence in a healthy outdoor activity. Sportsmanship, craftsmanship, self-reliance, discipline, application of creativity, and pragmatic thinking are only a few of the things that can be learned in model rocketry.

But, mostly, model rocketry is fun—as long as you follow the NAR Model Rocket Safety Code.

You'll never run out of things to do in model rocketry; it's an endless countdown, a hobby that you can grow up with and stay with for years because it offers a never-ending challenge to build better model rockets. It's the hobby of the final frontier.

Congratulations on getting hooked on the best aerospace hobby this side of Alpha Centauri!

2

GETTING STARTED

You can get started in model rocketry for less than $50, depending on the model and the equipment you select. This will set you up with a launch pad, an electric launch controller, a launching battery, a simple beginner's model rocket kit, and a few model rocket motors.

Most of the larger model rocket manufacturers (see Appendix I for names and addresses, correct as of this writing) make and sell starter sets that include all of the above equipment except the batteries that are purchased separately. You can also find a complete starter set that is ready to fly (RTF) containing a preassembled rocket. All you need to do is insert the batteries and rocket motors, and you're flight ready! You can buy all this equipment independently if you wish, but the prepackaged starter set will save you time and money. In any event, you will continue to use the launch pad and the electric launch controller in all your model rocket flying activities for a long time. So this big initial cost hits you only once. If you intend to pursue the hobby seriously, you should get the best launch pad and controller that you can afford. We'll talk a lot about batteries in a later chapter, but most electric launch controllers that come in starter sets use four AA alkaline batteries that should be purchased separately to ensure that they're fresh.

If you're handy with tools, you can make your own launch pad and electric launch controller. You'll find details about this in chapter 7.

Do not try to fly a model rocket without a launch pad!

Do not use any ignition system other than electric ignition with a battery and launch controller in accordance with the instructions of the manufacturer of the model rocket motor you are using.

Half a century of model rocket experience and more than sixty years of professional rocket activity have proven the validity of these two statements and the safety of these two necessary items. The few accidents that have occurred in model rocketry have been caused by attempts to take shortcuts in the requirements and safety rules for launch pads and electrical ignition. If

FIGURE 2-1   Two of the more commonly available ready-to-fly starter sets containing nearly everything you'll need to get started in model rocketry.

you're going to do something, do it right or don't do it at all. If you believe you can ignore all the technical experience and know-how and do it *your* way, no model rocketeer will sympathize when you have your inevitable accident. In fact, they'll be pretty angry with you for being a bad example when they work so hard to do it the right way, the safe way.

There are many different RTF rockets and sets available as shown in Figure 2-1. They can be a good way to get started with a minimal time investment. Very soon after launching your first RTF rocket, you'll be anxious to try your skills at assembling your first kit.

If you're a child, you'll probably need some help from a parent or other adult to put together your first model rocket starter set. In general, people under 10 years of age don't have the manual dexterity to build a model rocket and its launch equipment without help. *Don't fail to ask for help* if you get into a bind! The instructions that come with model rocket kits, motors, and equipment are probably the best instructions of any activity in the hobby field. The instructions with starter sets are written for people who have never seen the equipment and have probably never built any sort of model before.

Remember: *Read all instructions first! Then follow the instructions when you build and fly.*

When building, don't rush. Don't panic. Don't goof it up. Don't try to cover up the goof. Do things right the first time, and they'll work right for you.

I've helped thousands of model rocketeers get started. I know you're anxious to get your model finished and into the air for its first flight. But if you don't take the time to build your model rocket correctly and according to instructions, you're likely to be disappointed when you finally push the launch button.

Current beginners' models are marvels of top-notch product engineering based on careful studies of beginners' requirements and capabilities and on thorough field tests of the models in the hands of beginners. Just because a model happens to be a simple beginners' bird, don't get the idea that it won't perform. Many beginners' models have contest-winning performance because of their simplicity. I have set national model rocket performance records with beginners' model rockets.

Although it often takes months to build a contest-winning scale model, most people can assemble a beginners' model with an all-plastic tail fin unit in about an hour. The speed record for assembling such a model from the instant of box opening to the moment of launch is held by Greg Scinto of Stamford, Connecticut, who achieved this dubious honor by doing it in 11 minutes 56 seconds back in 1970. He would have done it in less time, but he accidentally glued his finger into the body tube at one point in the assembly. (I've seen a lot of funny things happen, and from time to time in this book I'll mention one of them to make a point or just to remind you that this is a serious book about having fun.)

Most model rockets, regardless of their details of construction, size, and performance capabilities, have the following parts and assemblies (see Figure 2-3):

1. A hollow plastic or solid balsa nose that fits onto the front of the hollow rocket body and that will come off.
2. A hollow, lightweight, thin-walled plastic or rolled paper body tube that is the main structural part of the model rocket airframe. This body tube holds the recovery device, the motor, and the fins.
3. A launch lug, which is a small tube like a soda straw (the first launch lugs *were* soda straws) that is attached to the exterior of the body tube. It slips over the launch rod of the launch pad, thus holding the model upright on the pad before launch and guiding the model during the launch phase of the flight.
4. A recovery device such as a plastic parachute or streamer that is packed inside the body tube and ejected forward from the body tube by the retro-thrust action of the model rocket motor at a predetermined time after launching.
5. The replaceable solid-propellant model rocket motor (most small model rocket motors are single-use expendable units, but some advanced reloadable motors now exist) and its associated thrust mount and retainer.
6. Fins made of balsa, plastic, cardboard, or fiberglass that are fastened on the rear end of the body tube and, like feathers on an arrow, keep the model traveling in a true and predictable flight path.

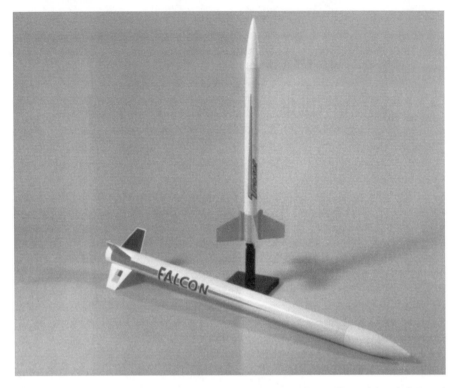

FIGURE 2-2   Two beginners' model rockets are the Quest Tracer (upright) and the Quest Falcon (horizontal). Both have plastic noses and fin units, color-coded parts, and no need for painting.

7. Recovery wadding to protect the recovery device from the gas of the retro-thrust action of the motor. Some kits have an internal baffle arrangement that eliminates the need for recovery wadding.
8. An electric igniter to start the solid-propellant model rocket motor.

The model rocket kit that you buy or that comes with your starter set will have all the necessary parts for you to assemble a model rocket with all of these features. It will be up to *you* to assemble them into a strong, lightweight, streamlined model that will slip through the air at speeds up to 300 miles per hour (mph) without breaking apart because of air resistance (drag). You will be assembling a model that flies faster than the fastest model airplane!

Don't let this shake you up. You can do it successfully *if you read and follow all instructions*! Millions of other people have done it. And today's beginners' models are a snap to build and fly. Just take your time, read the instructions, believe the instructions, and do it right the first time. Once your model rocket is in the air, you can't call it back to make a correction or to add something you forgot.

Many beginners' model rockets require no painting; all parts have been precolored. Others require painting, but many beginners don't paint them.

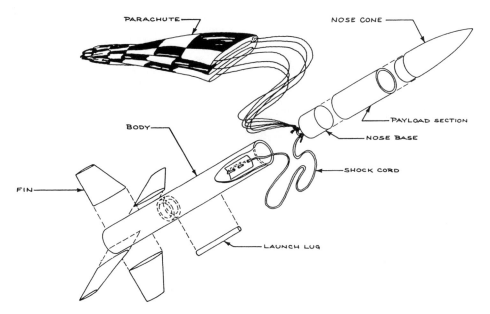

FIGURE 2-3   The basic parts of a typical model rocket are shown in this simplified disassembled view. The body tube is paper. The nose and fins are either
plastic or balsa. Other parts are paper or plastic. The motor is always a commercially available factory-made unit. This model has a payload-carrying section.

They think painting isn't required. However, paint a model rocket anyway for
a very good reason: If you paint it a bright color such as flat white, bright
orange, or fluorescent orange, you will be able to see it better in the air during
flight *and* in the tree in which it will inevitably land. You will also be able to
find it more easily on the ground should it happen to elude the clutches of the
rocket-eating trees that exist *everywhere* in the world. (Trees love model rockets. They eat them very slowly, savoring them for as long as several months.
You can watch parts of your favorite model disappear week by week as the
tree slowly devours it. This is a proven fact. At least, it is a well-known
hypothesis among experienced model rocketeers.)

Nearly all model rockets operate in the same basic manner, although performance can vary greatly due to differences in weight, size, shape, motor
power, and other factors that we'll discuss in detail later.

When you're finally ready for that first flight session, excitement will
reign supreme! But before you leave the house to go to the flying field, make
certain you have *all* of the following items:

1. A launch pad.
2. A launch rod.
3. An electric launch controller.
4. The safety key for the launch controller. Put the safety key on the end
   of a bright strip of cloth or ribbon so that you don't drop it and lose it.

5. Your model rocket (some people have forgotten them).
6. Three to six model rocket motors of the proper size and power for your model as recommended by the manufacturer.
7. An electric igniter for each model rocket motor plus two or three spares just in case.
8. A roll of paper tape such as masking tape.
9. A roll of cellophane tape.
10. The ignition battery or fresh batteries installed in the launch controller.
11. Plenty of recovery wadding.
12. Your parent or other adult.

Why this last item? Why the adult? Simply to ensure that you're successful. Flying a model rocket isn't difficult or complicated. But it requires that you do *everything* correctly in the proper sequence *before* you push the launch button. In that respect, it's just like a big expendable space rocket launch at Cape Canaveral. I highly recommend adult participation in the construction and flying of model rockets because I know too well from my own experience that a young model rocketeer's natural enthusiasm and excitement over flying a model rocket can cause him or her to overlook some important point in the countdown sequence. Many young people are perfectly competent to build and fly a model rocket all by themselves in complete safety. They may know more about it than an adult supervisor. However, the double-check feature of adult participation can often prevent mistakes made in haste and excitement.

The professionals use a double-check system at the Cape for the big rockets. Why should model rockets be any different in this respect from the big ones just because of their size?

Your next task is to choose a flying field. It should be away from major highways and freeways, power lines, and tall buildings. (How would *you* like to be cruising along the freeway at 55 mph and suddenly see a model rocket under a parachute descending right in front of your windshield?) The size of the flying field depends on how high you expect your model to fly. Check the NAR Model Rocket Safety Code for minimum dimensions. In general, for beginners' models propelled by Type A, Type B, or even Type C motors, a ground area about the size of a school athletic field is adequate on a calm day. On a windy day, the model will drift farther in the wind and will need some clear downwind area to land in . . . unless you want to feed the rocket-eating trees.

Set up your launch pad on the *upwind* side of the flying field so that the model will drift back across the field and land within its boundaries.

Important: After setting up your launch pad and electrical ignition system, *test it.* Hook up one of the spare igniters you brought (but do not install it in the motor). The igniter should not activate until you have put the safety key into the launch controller and pushed the launch button. If your system can't pass this simple test, you have problems. You'll have to troubleshoot the electrical system to find out what's wrong. If the igniter doesn't activate after

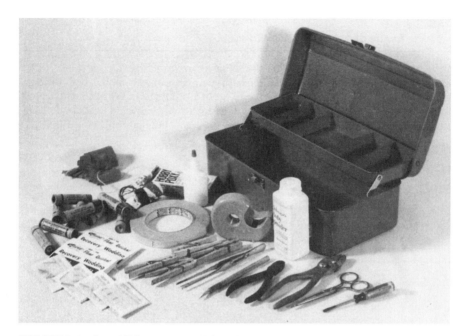

FIGURE 2-4   In addition to a model rocket, launch pad, controller, and battery, the fully prepared model rocketeer should have a range box with the tools, pieces of equipment, and other goodies that make flying easier and more fun.

you have inserted the safety key and pushed the launch button, it probably means your battery is too weak or dead. This rarely happens today when using the new igniters that can be activated by four AA alkaline penlight batteries except on a cold day, but it is a common problem with some of the older igniters. However, failure of the igniter to activate could also mean that you've got problems in the wiring if you made your own launch controller. Read chapters 6 and 7 on ignition and launching for details about troubleshooting, or read the troubleshooting instructions that came with the launch controller. If your launch system passes this simple prelaunch test, you can have confidence that the model will not take off while you're hooking it up—and that it *will* take off when you want it to.

Now *prep* or prepare your model for flight. Insert the recovery wadding if your model requires it. Fold up the recovery parachute or streamer so that it slides *easily* into the body tube. Stuff the shock cord down on top. Put on the nose, making sure that you don't jam the shock cord or parachute shroud lines between the body tube and the nose. The nose should be able to slide off easily.

For your first flight, use the *lowest power motor* recommended for your model. Usually, this will be a Type A motor. Check the model rocket kit instructions to confirm what motor type to use. Some larger beginners' models require a more powerful motor for the first flight. Nevertheless, use the *lowest* possible motor power. The first flight is a test flight. You will want to

TABLE 2
Altitude Range of Average Model Rockets
_____

A small model rocket (0.787-inch or 20-millimeter diameter) and a large model rocket (1.378-inch or 35-millimeter diameter), both with the same shape but with different empty weights, were used to calculate the range of altitudes to be expected with a given motor type. Air resistance (drag) is accounted for (drag coefficient = 0.75). Calculations were made with the RASP-93 computer program shown in Appendix III.

Motor Type	No-Motor Weight (ounces)	Altitude Range (feet)
A6	1	224–337
	2	131–153
	3	75–80
B6	1	407–720
	2	316–435
	3	216–255
	4	145–158
C6	1	710–1355
	2	655–1081
	3	540–758
	4	416–515
	5	311–354
	6	230–249

get the model back so that you can fly it again. (Anyone can ram a Type C or larger motor into a model and lose it on the first flight. That's no outstanding achievement.)

Make certain that the motor cannot slide out of the rear end of the model. Most models today have a motor clip that holds the motor firmly in the body tube. If your model doesn't have a motor clip, wrap the motor with tape until you have to force it into place. Be careful! Too much tape and too tight a fit may cause you to buckle the body tube when trying to get the motor in or out later.

Install the igniter in the nozzle of the model rocket motor, carefully following the instructions that come with the motor package. Now your model rocket is ready for its preflight safety inspection, given by an adult who's supervising the flight activities. Why bother? Simply because it's too late to correct a mistake after you've launched the model. Once it's on its way skyward, you can't possibly stop it to fix a mistake or do something you've forgotten.

During all preflight times, keep the safety key to the launch controller in your pocket or hand. This ensures that nobody but you will be able to switch the electricity into the model rocket motor igniter. You will be sure that the circuit is safe to hook up the igniter.

Slide the model down the launch rod. The launch lug should slip easily over the rod. The function of the launch rod should be obvious to you by now.

It supports the model during hookup. And it provides the initial guidance of the model after ignition while the model's airspeed is still too low for the fins to stabilize the model.

Clear the launch area. Hook up the igniter. Make sure the clips do not touch each other and that they are not touching the metal jet deflector. Get in the habit of keeping your fingers out of the direct line of the exhaust jet while you're doing this because that good habit could prevent possible burn injury later in your rocketry career if anything is wrong with the firing system. Also, never look down on the model as it sits on the launch pad with the igniter hooked up and all ready to fly.

Now move back to the launch controller at least 15 feet from the launch pad (30 feet for Type E and larger motors). Keep everyone back the same distance or more. Check the flying field. If the launch area and sky above are clear and the situation is safe for launching, the adult safety officer should give you a "safety go" for launch. Insert the safety key in the launch controller. The ignition continuity light should come on, telling you that you have a hot circuit ready to launch. (If the light doesn't come on, pull the safety key and begin troubleshooting operations.)

Always give a 5-second countdown in a loud voice: "5 . . . 4 . . . 3 . . . 2 . . . 1 . . . *Start!*"

When you press the launch button on the controller, the electric current will flow from the battery to the igniter in the model rocket motor. The electrical resistance of the igniter will cause it to glow red-hot instantly (or less than instantly if you have a weak battery). This ignites the solid propellant in the model rocket motor.

The solid propellant begins to burn, creating a large volume of hot gas that rushes out the rocket nozzle at more than twice the speed of sound, producing a thrust force that accelerates the model on its way. The model will lift off very quickly. It will reach the end of the launch rod in about $2/10$ second and be traveling at 30 mph or more as it takes to the air.

Powered flight lasts for 1 second or less. During this time, all the solid propellant in the motor is used up. At this "burnout" point in the flight, the model will be 50 to 200 feet in the air and traveling at a speed between 100 and 300 mph straight up.

Now coasting flight begins. The model trades speed for altitude. The end of burning of the solid propellant in the motor has started a slow-burning time delay charge in the motor. This time delay produces no thrust and allows the model to coast upward, delaying the ejection of the recovery device until maximum altitude has been attained.

At or near apogee—the maximum distance from the earth—and at a time determined by the choice of a model rocket motor with the proper time delay built into it, the recovery ejection charge in the motor activates. This produces a retro-thrust puff of gas that pressurizes the body tube, forcing the wadding and recovery device forward and dislodging the nose. The recovery device deploys into the air and slows down the model. The model then returns slowly and gently to the ground. Since all parts are tied together by the shock cord,

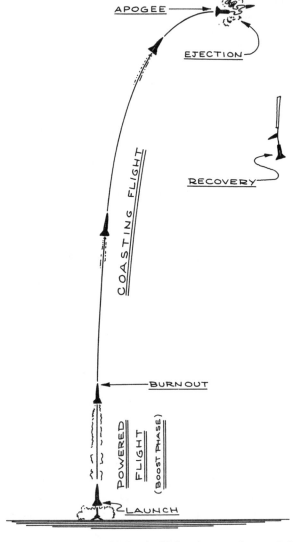

FIGURE 2-5 This sketch shows the basic flight phases of a model rocket. This is the sort of performance you should expect of your first model rocket—and every one you build thereafter, too.

the entire model lands together where it can be recovered easily—unless it has (naturally) landed in the one and only tree in sight.

If the model lands in an electric power line, forget it. Don't *ever* try to get a model rocket out of a power line. You can buy a new model rocket kit for a few dollars. You cannot buy back your life.

With your model rocket happily in hand, return to the launch area. Check the model for damage such as broken fins or cracked parts. Take out the

FIGURE 2-6   Liftoff! The moment of truth arrives as you push the ignition button and your model rocket starts to streak aloft from its launch pad.

expended model rocket motor casing and put it in your field box. Take it home and throw it away. Don't give it to the little kids who will immediately come out of the bushes within minutes after your first launch. These kids may try to stuff the used casing full of match heads and get hurt. Repack the recovery device, install a new model rocket motor and igniter, and you're ready to fly again.

I have seen model rockets that have made more than 100 flights—but they didn't make 100 flights in one afternoon. You'll be doing well to get off 3 or 4 flights in your first flight session. Try some model rocket motors of higher power and notice the difference in performance of your model. Experiment with tilting the launch rod a few degrees away from vertical (but never more than 30 degrees).

TABLE 3
Model Rocket Skill Levels

Most model rocket manufacturers have adopted a universal skill level coding of their kits. Skill level 1 is the easiest to assemble. With each increasing skill level number, the model becomes more difficult to assemble and fly. Once you've mastered a particular skill level, you are ready to move on to the next one! (E2X is a registered trademark of Estes Industries.)

Ready to Fly (RTF). No paint, glue, or modeling skills required. Rocket comes assembled and is ready for flight in just minutes.

E2X or Quick-Kit. No paint or special tools needed. Kits contain parts that are colored and easy to assemble. Assembly takes as little as 1 hour.

Skill Level 1. Requires some measuring, gluing, sanding, and painting. Assembly takes at least an afternoon.

Skill Level 2. Requires some basic beginner skills in model rocket construction and finishing. Assembly time may take a complete day.

Skill Level 3. Requires moderate skills in model rocket construction and finishing. Assembly may take a couple of days.

Skill Level 4. The most advanced level of kit. Requires a high degree of model rocket construction and finishing experience. Assembly can take several days depending on the detail required.

Once you have put in your first afternoon of flying model rockets, you're ready to go on to bigger and better models. I'm certain that you'll be hooked on model rocketry at this point. Supervising adults may be hooked, too. I've seen many parents who grumpily went along to the flying field for the first flight because their young rocketeers insisted on doing things the right way, and I've seen those parents come home afterward far more excited about building and flying model rockets than their offspring.

Model rocketry isn't just a kid's hobby. More than half the people flying model rockets in the United States today are adults. Some of them started in model rocketry when they were children themselves. After drifting away from the hobby in high school, they came back when they had children of their own. Or they returned because they had fond memories of the fun and friends of model rocketry. These are the BARs (born again rocketeers).

All model rocketeers have one thing in common: we all remember our first model rocket flight. It's an important experience. People remember it even after flying thousands of model rockets. I can still remember in detail my first model rocket flight on a chilly February morning in a cotton field in Las Cruces, New Mexico, in 1957. Today, many thousands of flights later, I get the same kick out of pushing that launch button and watching one of my creations lift off into the sky.

Sure, it won't get to the moon—but I can imagine that it does!

3

TOOLS AND TECHNIQUES
IN THE WORKSHOP

In working with model rockets, common sense is required. If you don't have it to start with, you develop it very quickly. You can develop your manual dexterity and your ability to work with tools as well.

The most important rule for becoming a successful model rocketeer is simple:

Read and follow all instructions completely.

The second important rule is:

If at first you don't succeed, go back to rule number 1.

I cannot emphasize these two rules too strongly. The biggest mistake made by novice model rocketeers (and even by some expert model rocketeers who think they know better but don't) is failing to read and follow instructions. As an indication of what can happen when a model rocketeer gets rushed or becomes a victim of "I know it all," I have actually seen the following gross mistakes in beginners' model rockets:

1. A nose glued to the body tube, so the nose could never come off for the recovery device to deploy.
2. A motor mount glued in backward, so it was impossible to insert a model rocket motor of any type into the model.
3. A solid bulkhead glued across the inside of the body tube, so no ejection charge gas could eject the recovery device.
4. Recovery wadding glued into the model.
5. Balsa fins that were cut out with a pair of scissors. "But, Mr. Stine, it makes such a nice, fuzzy edge for gluing!"
6. A model rocket motor firmly glued into the model.

SHOP RULE

If at first you don't succeed,

Try following the instructions.

—The Old Rocketeer

FIGURE 3-1   The basic rule, and don't forget it! Photocopy this and post it in your workshop.

7. An endless, disheartening parade of model rockets with crooked fins, fins that were too small, fins glued on the nose of the model (a definite no-no), no launch lug, launch lug mashed flat, no motor mount, shock cord not glued in, and (yes!) body tube crimped or bent.

Such models are garbage birds good only for tossing into the trash can. I know that *you* can do better!

Once you have built and flown your first beginners' model, you should go on to more difficult kits and eventually to designing your own models. First, give up every notion you may have had about rockets. Take that design for the four-stage radio-carrying mile-high supersonic rocket and tuck it away in your notebook; you aren't ready to build it yet. You have a lot to learn, and you'll have a lot of fun learning it. Later, when you look with greater knowledge and experience at that early, primitive design, you will either laugh at yourself or discover that the dream design needs a lot of changes based on what you've learned. This is true even if you're an experienced professional rocketeer working for NASA or the U.S. Air Force who has taken up model rocketry as a hobby, or if you're already a champion airplane modeler who's decided to try something different.

We do many things differently in model rocketry, and we do them for a definite reason and for a specific purpose. Sometimes we've learned about them the hard way. There's no need for you to repeat the past mistakes of others. In this book, you'll learn *why* we do things the way we do. You'll be able to see why for yourself in your own workshop and out on the flying field.

WORKSHOP

First, you must have a place to work. Beginners' models and some other simple kits don't require much of a workshop. But if you want to progress beyond simple models, you'll have to develop a workshop. This is a place where you can keep your tools, your model rocketry supplies (spare noses, body tubes, balsa, and other things that you'll accumulate or salvage from broken models), your models as you're building them, and your models once they're finished.

A card table in your bedroom is all that you may need as you're beginning. But balsa dust, wood chips, paint overspray, and other debris accumulate while you're building model rockets. You really need a place that will keep this stuff from spreading itself thinly all over the rest of the house to the dismay of your family. In addition, modeling paints and glues tend to be somewhat aromatic. In other words, to a person who isn't a modeler, they stink.

If you can manage to commandeer a corner of the basement as a workshop, be sure there are windows that you can open to provide ventilation. The days of glue sniffing are over because glue manufacturers have incorporated into the glue a substance that makes you violently sick to your stomach if you inhale too much glue vapor. The vapors of drying paints aren't good for you, either. So make sure you have adequate ventilation. Don't short yourself on fresh air.

Garages make excellent workshops in warm climates. If you live in a cold climate and can get a garage as a workshop, it should be heated because it's difficult to do precision modeling work with cold fingers or while wearing gloves and other cold-weather clothing. Glues and paints are also affected by freezing temperatures. However, if the garage has a space heater, keep paints, glues, model rocket motors, and spray paint cans away from that heater unless you want to put the garage into orbit.

Wherever you manage to set up your workshop, try to keep it in some semblance of order and cleanliness, impossible as that may be at times. (Some people believe that cleanliness is next to godliness, but for a model rocketeer

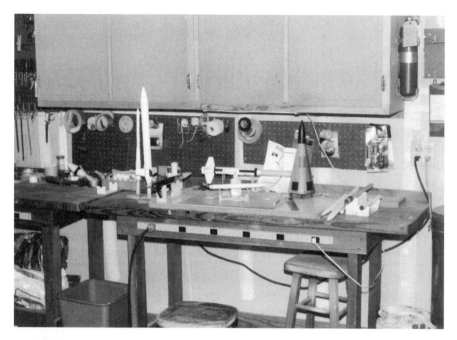

FIGURE 3-2   A well-lit, well-equipped workshop is essential for good modeling. This shop runs across one end of a two-car garage.

cleanliness is next to impossible.) This keeps you from tracking balsa dust into the house. It permits you to find things more quickly when you're in that panic rush to complete the model the night before a major contest. It also helps in finding the little bitty part that obeys Paul Harvey's Law: A dropped part will always roll into the most inaccessible part of the workshop. Mostly, however, a neat and clean workshop makes it easier for you to get along with the other nonmodeling members of your family.

SIMPLE TOOLS

Your most powerful tool is a *notebook* in which you can write down your ideas, make sketches, note how you did things so that you can remember them better in the future, and file technical reports and other papers. I have more than a dozen loose-leaf notebooks, each of them crammed with information gathered over the years. They are the most important books in my library. *Handbook of Model Rocketry* has been compiled and updated from the information I've tucked away in these notebooks. Scientists and engineers always keep notebooks and logbooks for their ideas, progress notes, reports, and other data. Model rocketry is a scientific and technical hobby, so it will pay for you to start keeping a model rocketry notebook from the very start. There is a tremendous amount of information and data available in model rocketry, far

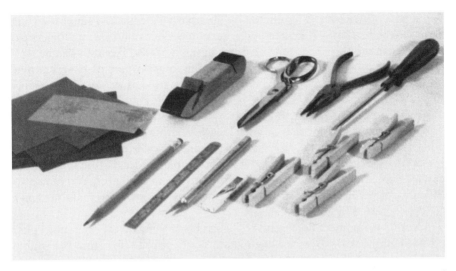

FIGURE 3-3   Some of the simple tools you'll use most to build all sorts of model rockets.

more than in most hobbies. You won't want to lose it once you've got it, so put it away where you can find it again—in your notebook. Your notebook will keep you from reinventing the wheel, which is duplicating somebody else's mistake that you should have read about, noted, and filed away to keep you from doing it, too.

Because you can't put a model rocket together with your bare hands, you'll need a few simple tools and you'll have to learn how to use them. You may already have some model rocketry tools. All of them can be purchased quite inexpensively at your hobby shop or hardware store. (Some model rocketeers also happen to be tool collectors; the tool department of a hobby shop is like a candy store for these people.)

The use of tools is one of the many things that sets human beings above the beasts of the jungle. So get good tools to start with, treat them properly, and use them safely.

You'll need a *modeling knife*. You can buy a modeling knife at a hobby shop for a couple of dollars, and it will last for years. It has a slim metal handle with a holder or chuck at one end to grip a special, extra-sharp replaceable blade that will cut balsa, plastic, and paper. A typical modeling knife is called an X-Acto knife. One of the best knives for modeling is the X-Acto No. 1 knife. I like to use it with the X-Acto No. 11 blade. Buy a package of extra blades for your knife so that you'll always have a sharp blade. When the blade gets dull and won't cut neatly—this happens quickly when cutting paper— remove it from the handle and install a new one.

Note: When you use a modeling knife, be careful. It cuts fingers more easily than it cuts anything else.

A *pair of tweezers* is handy for holding parts that are too small for your fingers to hold. Thus, they become a very fine extension of your fingertips. And you can use them to reach and grasp in small places. Tweezers can be purchased in a hobby store. You may also want to get a pair of tweezers that *unclamp* as you squeeze them so that they can be used as a long, thin clamp.

A sharp *pencil* or a *ballpoint pen* is a must for taking notes, marking parts, etc. To keep it from "walking away" from your workbench, put it on a leash. Tie or glue one end of a 24-inch length of string around it and attach the other end to your workbench with a tack, nail, or staple.

Scissors are useful for cutting out paper templates, decals, and other paper parts. As mentioned before, please don't use them to cut balsa. You may also have to attach your scissors to your workbench with a string so that they don't get appropriated by someone else.

Needle-nose pliers are useful for assembling things with nuts and bolts, for holding parts, and for bending metal. Again, your hobby store is a source for these.

A *small screwdriver,* both slotted and Phillips-head, is a model rocketeer's friend in the shop and in the flying kit you take to the flying field.

A metal 6-inch or 12-inch *rule* will be used to measure and to guide you in cutting straight lines with a modeling knife. Don't get a plastic or wood rule; you'll end up cutting its edge and will therefore make balsa cuts that are as crooked as a mountain highway. To cut a straight line with a modeling knife and a rule, merely lay the steel rule down on the balsa or paper and run the knife along the rule edge.

If you can find a metal rule with both American system inches and metric system millimeters or centimeters on it, you'll discover how easy it is to measure things in the metric system. It's easier to work in the metric system when building model rockets. A millimeter, for example, is $^{40}/_{1,000}$ of an inch (0.040 inch), or roughly $^{25}/_{64}$ inch. It is much easier to work in the round numbers of the metric system than in the fractions of an inch of the American system. For example, it's easier to work with 15 millimeters than with its close equivalent, $^{19}/_{32}$ inch. Model rocketry is technically on the international metric system and has been since 1962. However, many measurements are still given in the American system of inches, pounds, etc., because most Americans are familiar with these units. In model rocketry, you can become quickly and easily acquainted with the metric system.

A good stock of *sandpaper* in various grits should be kept at hand. It is used for shaping and smoothing. It comes in various grades ranging from very coarse to very fine. For model rocket work, you can buy large sheets of sandpaper at a hardware store at very low cost. Cut these big sheets into little pieces about an inch (25 millimeters) square using a pair of scissors. Get sheets of No. 200, No. 320, and No. 400 wet-or-dry sandpaper. No. 200 is useful for fast shaping, No. 320 is good for all-around stuff, and No. 400 is great for final smoothing.

You can buy a *sandpaper block* at the hobby store, or you can make one by taking a convenient-size wood block and tacking some No. 200 sandpaper

TABLE 4
Metric Conversion Chart

From	To	Multiply by
Centimeters	Inches	0.3937
Centimeters	Feet	0.0328
Inches	Centimeters	2.54
Inches	Millimeters	25.4
Inches	Meters	0.0254
Meters	Inches	39.370
Meters	Feet	3.2808
Meters	Yards	1.0936
Feet	Centimeters	30.48
Feet	Meters	0.3048
Feet	Kilometers	0.0003048
Grams	Ounces	0.03527
Grams	Pounds	0.0022046
Kilograms	Ounces	35.2739
Kilograms	Pounds	2.2046
Ounces	Pounds	0.0625
Ounces	Grams	28.349
Ounces	Kilograms	0.0283949
Pounds	Grams	435.592
Pounds	Kilograms	0.45359
Newtons	Pounds-force	4.45
Pounds-force	Newtons	0.2247
Pounds-weight	Pounds-mass	0.03105
Pounds-mass	Pounds-force	32.27
Miles/hour	Feet/second	1.467
Feet/second	Miles/hour	0.6818
Meters/second	Feet/second	3.281
Feet/second	Meters/second	0.3048

to it. The wood block provides a flat, firm base for the sandpaper and permits you to shape flats, curves, and weird shapes more easily.

You'll need some *clamps* to hold parts together while glue is drying, to hold parts generally, and to use around your launch pad on the flying field. The finest modeling clamp is the old-fashioned spring clothespin. It can also be sawed, drilled, carved, sanded, painted, glued, and made into many needed items.

A *cutting mat* is used to protect your tabletop or workbench from cuts made using your modeling knife. These unique mats have a special polymer surface that "heals" itself after you make an incision with a modeling knife. They are available in a variety of sizes (12 by 18 inches is usually about right for most workbenches).

These are the basic tools of the model rocketeer. I've built a lot of good model rockets in a one-room studio apartment or in a motel room with no tools other than these.

OTHER HELPFUL TOOLS

Probably the most ubiquitous power tool in American homes is the $^1\!/_4$-inch or $^3\!/_8$-inch *electric drill*. This is a very handy model rocket tool if you use some ingenuity to set it up in various ways. Sure, it will always drill holes if you put a drill bit in the chuck. But by using the drill press accessory available where you bought the drill, you can turn a simple electric drill into a very effective and accurate drill press, buffer, polisher, and grinder. With a horizontal holder, the electric drill becomes a useful miniature lathe for turning model rocket parts out of balsa. Many modern electric drills have various speed control and reverse features; they shouldn't be called electric drills because they are, in reality, portable rotary electric power tools. Think of your electric drill in these general terms, and you can find a lot of uses for it.

If you have a small wood *lathe* or can get access to one in a school shop, you're really in the model rocket–building business! Such a lathe can be used to make many model rocket parts. Although the Austrian Unimat lathe is expensive, it is extremely accurate and will last for a long time. I bought one in 1957 and have used it in my model rocket work ever since. It has paid for itself many times over because I've been able to make a lot of special model rocket parts with it instead of buying them. Other small, inexpensive wood lathes are available. Of all the power tools in a model rocket workshop, the lathe is perhaps the most used and most helpful.

Other handy tools include:

A claw hammer

A small Ball Pein hammer

Various sizes of screwdrivers, slot and Phillips-head

Side cutters

Diagonal cutters

Soldering iron

Model maker's tee-head pins

A bench vise

Paper clips

JIGS AND FIXTURES

All sorts of handy little jigs and fixtures have been developed by model rocketeers over the years. The following are some that I've found to be useful.

Cradle stand—How do you set a model rocket down on the workbench while you're working on it? Answer: Do as NASA rocket engineers do and use a cradle. A typical cradle can be cut from balsa sheet, bristol board, poster board, or other stiff cardboard as shown in Figure 3-4. If you have a model that is shorter or longer than the cradle shown in Figure 3-4, you can make a

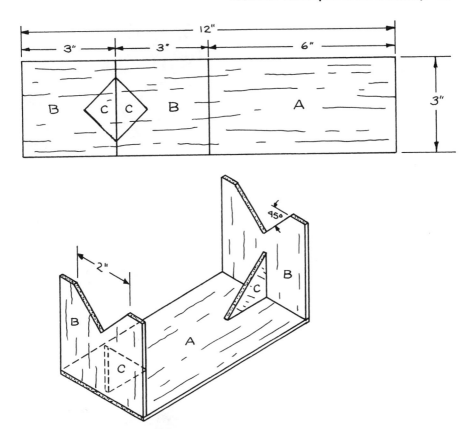

FIGURE 3-4   A cradle stand to hold a model rocket in a horizontal position while working on it can be cut from sheet balsa or stiff cardboard as shown. Dimensions can vary to suit the size of the model rocket.

cradle longer or shorter than this. You can make it bigger for large models. I have a series of cradles ranging from 3 inches to 12 inches long and some that will handle tubes up to 4 inches in diameter. Sometimes I paint one if it's to be used to display a model. You'll find that a cradle is one of the handiest fixtures in the shop.

Spike—Wendell H. Stickney made the first rocket spike in 1961, and I still have it and use it. A spike holds a model rocket vertically during construction or display. The easiest way to make a spike is to glue an expended motor casing to a block of wood. However, you can make a better one with a lathe to help out. A typical spike is shown in Figure 3-5. It's longer than a regular motor casing and will hold a model rocket high enough to clear any swept-back fins that extend beyond the rear end of the body tube. A spike will increase your workbench space because it holds a model vertically.

Spike row—The obvious and logical extension of the spike is several spikes fastened to a long board (see Figure 3-6). This is useful not only in the

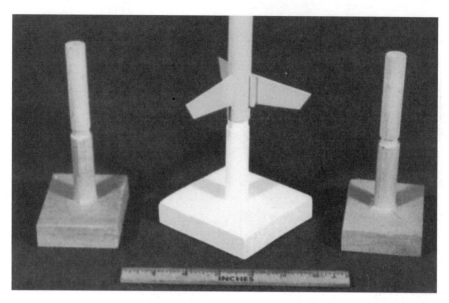

FIGURE 3-5   The workshop spike can be made from wood or plastic. It is designed to hold the model vertically.

FIGURE 3-6   A spike row is a line of spike dowels on a single board for holding several models vertically.

FIGURE 3-7 A paper painting wand is made from a section of newspaper rolled up and stuck into the rear end of the body tube. It keeps you from painting your hand when spray painting a model rocket.

shop or for an impressive display but is handy when you have to carry a lot of models around. It's as easy to make as a single spike. Most of my spikes and spike rows are made to handle model rockets built for standard 18-millimeter model rocket motors. I've built others to handle 24-millimeter motors, too.

Paper wand—How do you paint a model rocket without getting paint overspray all over the hand that's holding the model? Make a paper wand by rolling up a sheet of newspaper and fitting one end of it into the rear end of the model. Hold on to the other end. Spray only at the model end, of course. The flimsy newspaper becomes quite strong after you've rolled it into a wand. Figure 3-7 shows a model being spray-painted using a paper wand.

Rotating spike—For painting models with an airbrush, spray gun, or spray can, I built a spike mounted horizontally on a support that would permit the spike to be rotated horizontally (see Figure 3-8). The spike is slipped into the model's motor mount tube. The model is thus held horizontally and can be rotated. This permits easy spray painting of the entire model. Making a rotating spike requires some work with a lathe and drill press, but it isn't so complicated that you can't finish it in a couple of hours as a quickie project.

Fin alignment angle—How do you draw a line accurately along a model rocket body tube so that you can correctly locate the fins? The answer is so simple you may not believe it. Go to a hobby shop or hardware store and buy

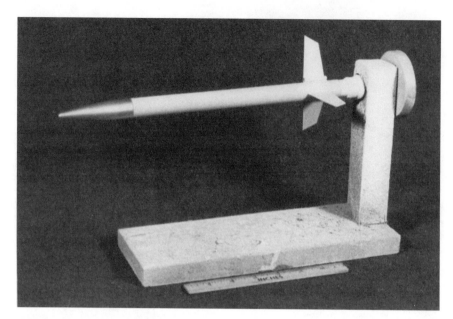

FIGURE 3-8 A rotating spike holds a model rocket horizontally so that it can be turned to spray paint all sides of it.

FIGURE 3-9 Art Rose developed this extremely accurate fin alignment fixture. The body tube slides over a variety of different mandrel sizes.

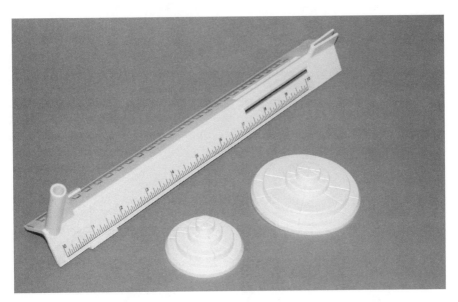

FIGURE 3-10 This set of tube marking guides is available from Estes.

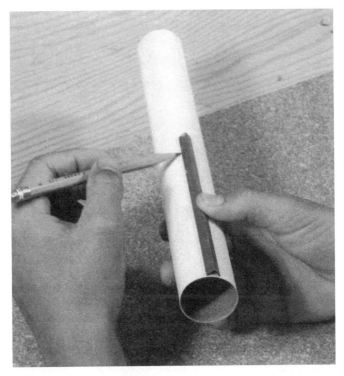

FIGURE 3-11 A small metal angle laid against the side of a body tube permits a straight line to be drawn down the tube for positioning the fins.

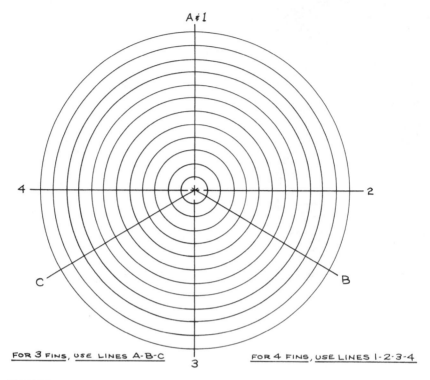

A ≠ l

4 ——— 2

C

B

3

FOR 3 FINS, USE LINES A-B-C

FOR 4 FINS, USE LINES 1-2-3-4

FIGURE 3-12   With a fin positioning drawing such as this, fins can be properly located on the end of a body tube.

a small piece of metal angle about 6 inches along. You should get ½-inch angle or smaller. If you can find some brass angle used in model railroading or for slot car bodies, it's perfect. Simply lay the angle down along the body tube as shown in Figure 3-11. Use one edge of the angle as a ruler to draw a straight line along the body tube. The line will be perfectly aligned with the long axis of the tube. A very useful tube-making guide is also available as shown in Figure 3-10.

Fin positioner—In the history of model rocketry, thousands of different fin assembly jigs have been designed to help the rocketeer get the fins glued on straight and true. The simplest one I know of is the Kuhn fin jig shown in Figure 3-13. Go to a hardware store or lumberyard and get a piece of wood angle trim. Cut it to a length of about 6 inches. With a saw, cut a slot at the apex of the angle as shown. Make the slot about ⅟₁₆ inch wide for normal fins; you can make another jig with a slot ³⁄₃₂ inch wide or ⅛ inch wide for heavier fins. Put the angle on the body tube as shown and hold in place with two rubber bands around the body tube. Insert the fin through the slot and the jig will hold the fin in perfect position while the glue dries. The design of the Kuhn jig eliminates the worry about gluing the jig to the model in the process. If you

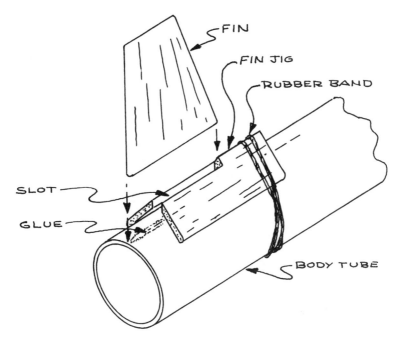

FIGURE 3-13   The Kuhn fin jig can be made from a piece of wood angle (household interior trim material). It will hold a fin straight on a body tube while the glue dries.

want the ultimate in precision fin alignment, you can invest in a device like the one shown in Figure 3-9.

Cotton swab—Often you'll need to put a bead of glue around the inside of a body tube in order to glue a motor mount in place. The easiest way to do this is with a tool swiped from the bathroom medicine cabinet and widely known by the trade name Q-tip. It is a wooden or paper stick with a ball of cotton on one or both ends. You can easily make one using a small wood dowel and cotton. Or you can buy a box of them at the drugstore. Mark on the stick the distance inside the body tube for glue application. Put glue on the cotton swab and carefully insert it into the body tube, being careful not to get glue on the inside walls until the swab has reached the proper insertion depth. Then swab the glue around the inside of the tube (see Figure 3-14). It may take several applications before you get a good band of glue in there. Save the Q-tip because you can use it many times. In fact, it gets better and better as you continue to use it because the hard glob of dried glue that builds up on the cotton makes it easier to use.

4¹⁄₂-inch craft sticks (Popsicle sticks) and index cards—I always mix my epoxy on a clean index card using a craft stick. They are both very cheap and can just be thrown away after you use the epoxy.

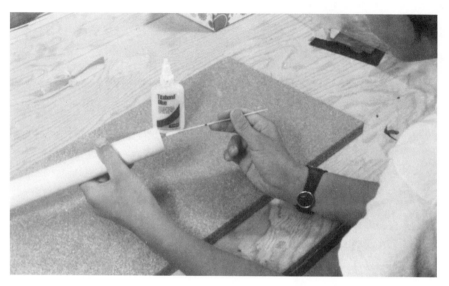

FIGURE 3-14   A cotton swab can be used to apply glue to the inside of a body tube. Mark the swab stick to indicate how far into the tube to apply the glue.

Tissues or paper towels—I always wind up with extra glue on my finger, especially after smoothing out a glue fillet on a fin/body tube joint. Having tissues or paper towels handy is a good idea.

Many other special tools, jigs, and fixtures have been developed by model rocketeers. I'll probably hear very shortly about all of those I haven't mentioned here. But I've told you about those that I've tested and found to be most suitable and useful in my model rocket workshop.

GLUES

We used to have only one kind of glue to use for building model rockets: model airplane glue. But the field of bonding technology has progressed, and many kinds of adhesives are available today. We call them glues, although in the true technical sense they are bonding agents. The wide variety of bonding agents gives the model rocketeer a lot of choices. In fact, the way to start a good argument among model rocketeers is to talk about your favorite type of glue. Everyone seems to have a different favorite.

The materials used in model rocketry are not difficult to join by bonding or gluing. But you must use the proper bonding agent for the material(s) you are joining. Porous materials such as paper, cardboard, and balsa require a different kind of bonding agent than nonporous materials such as plastic. The following information on glues and bonding agents has been accumulated by me during years of building and from correspondence (and arguments) with other aerospace modelers from all over the world.

FIGURE 3-15   A wide variety of glues and bonding agents can be used in constructing model rockets; some are intended only for special applications.

Model airplane glue such as Testors, Ambroid, and others is rarely used in model rocketry today.

White glue such as Elmer's Glue-All is casein-based. It is made from milk by-products and is water soluble before it sets up and hardens. It is exceptionally strong for use with paper, wood, and other porous materials, but it will not work on plastics. Its major disadvantage is its long drying time. It may take an hour to set and overnight to dry completely. You may also have some trouble applying some kinds of modeling paint over white glue layers. The craft industry has developed some fine white glues that are useful. I've been using Aleene's Original "Tacky" Glue because it is thick, dries reasonably fast, does not shrink as it dries, and dries clear.

Aliphatic resin glue such as Franklin Titebond looks like a brown-yellow version of white glue, but it has a different chemistry. This type of glue is very strong. It will set up and dry faster than most white glues. It is nearly transparent when dry. Most kinds of paint will go over it nicely. It will bond porous materials but not plastic.

Contact cement includes rubber cement, Goodyear Pliobond, and Weldwood. To use these bonding agents, you must be quick and accurate. Apply a coating to both surfaces to be joined, and allow this first coat to set for a

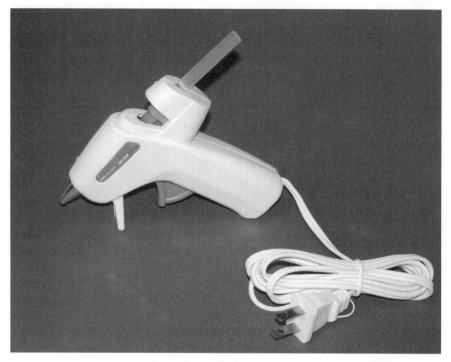

FIGURE 3-16   A craft hot melt glue gun is often used by teachers or youth group leaders during a group building session. The hot melt glue sets fast but gives students ample time to check their final fin alignment before setting.

few minutes. Then carefully join the surfaces. Contact cements bond firmly *at once* upon contact, so you have no leeway in moving parts around to align them. Contact cements will usually bond anything to anything else, but it's always a good idea to test first using a couple of pieces of scrap material.

Epoxy is now commonly used in modeling. Epoxies are strong, fast-curing, and universal in their bonding capabilities. They'll bond almost anything to anything else except Teflon and some kinds of polyvinylchloride (PVC). Basically, epoxy is a plastic technically known as *thermosetting*. Epoxy comes in two parts: a resin and a hardener. When you're ready to use epoxy, mix an equal amount of resin and hardener (or follow instructions for the specific epoxy because they aren't all alike). Mix only the amount you'll need in the next few minutes. Some epoxy types will cure or set in 60 seconds, but they do not have the strength of epoxies that take longer to cure. The strongest epoxies take about an hour to cure. Mix only what you need for a particular joint because epoxy has a limited working life once it is mixed ranging from 30 seconds for the fast stuff to up to 30 minutes for the long-term curing types.

Cyanoacrylate (CA) has revolutionized model building. CA is generically called "hot stuff," although this is the trade name of one of the more popular model-building CAs. This bonding agent is a liquid that acts almost instantly and will bond anything to anything. It cures or sets up immediately by application of pressure between the two parts that are to be joined. You can apply model rocket fins and other external parts very rapidly with CA. It's also great for quick repairs on the flying field. *Warning:* CA will cause immediate bonding of human skin to anything! Don't get it on your fingers, in your eyes, in your mouth, or anywhere else on your body. "Debonders" are available, and you should have a bottle handy if you use CA. However, if you bond yourself to something, including your own body, medical surgery may be required to separate you from yourself or your model. CA is great stuff, but be careful working with it. Some CA types don't bond paper very well. Test before using CA for joining parts that are subjected to high stresses in flight.

Cements or glues for plastics work differently than glues mentioned thus far. Plastic cements actually soften and melt the plastic material so that when two plastic parts are pressed together, the plastic flows together and welds.

Professional makers of plastic models use acetone or methyl-ethyl-ketone (MEK) that they buy in large cans holding a gallon or more. You can go this route if you are building 15,674 plastic models in the next few weeks. Otherwise, buy MEK and liquid plastic cements in small bottles from the hobby shop and keep the caps on tight when you're not using them. They're all highly volatile and will evaporate quickly if left open to the air.

Plastic cements also come in tubes like some kinds of wood cements. This sort of plastic cement is actually plastic material with lots of solvent in it to make it very soft and viscous. When the solvent evaporates, the cement becomes solid plastic. When you apply it to a plastic piece that will be joined to another plastic piece, the glue forms a bridge of plastic that unites the two parts. Most hobby plastic cements are intended to bond polystyrene or ABS, the types of plastic used in most plastic model kits. They don't work very well with other plastics such as PVC. Special glues are now available that will bond almost any plastic (except Teflon, which doesn't want to bond or react with *anything*).

When in doubt about the compatibility of plastics and certain glues, test them first. Some plastic glues will melt plastic, and this may ruin a plastic part.

In fact, testing glues with materials is a good idea, given the wide variety of different materials and bonding agents now available, with more coming along all the time. If you don't know whether the glue you have will work, make some tests using scraps of material. Glue them together and see what happens. This doesn't take very long to do, and it may save you some grief such as ruining your model during construction or having it come apart in flight at 300 mph.

There is a humorous side to bonding techniques. In aviation, designers have to contend with the speed of sound. In physics, it's the speed of light. In model rocketry, a similar "barrier" speed exists known as the speed of balsa. This is the speed at which balsa construction fails in flight. It is a variable constant because it depends totally on the individual model builder. Some people consistently build model rockets that have a very low value of the speed of balsa because they don't know how to select and use bonding agents. Or they don't follow instructions.

You can increase the speed of balsa tremendously by making a proper glue joint. Having a good glue is only half the battle. Knowing how to use it properly is the other half. Almost 90 percent of the people who make model rockets don't know how to make a good glue joint, even though the instructions tell them how to do it. This is probably because they think the correct method is only a tactic to get them to buy and use more glue. Not so.

If you want to make a good glue joint that is stronger than paper or balsa, follow these simple instructions for *all* kinds of bonding agents:

When gluing together porous materials such as paper or wood, use a *double-glue joint*. Coat both surfaces with a layer of glue or bonding agent and let both surfaces dry. Then coat both surfaces again and join them together. The first coat of glue on both surfaces penetrates the pores of the material. The second glue coat is then free to join with the first coat and with the second coat on the other surface. A double-glue joint will be so strong that the materials will break or tear before the glue joint turns loose. Try it. It will surprise you. And you'll never use any other kind of glue joint again.

When gluing two pieces of plastic or other nonporous material, use a variation of the double-glue joint. Coat *both surfaces* with the bonding agent. Since the surfaces don't have pores, you don't need to apply two separate coats of glue to the surfaces. But the cement will soften each surface or adhere to each surface so that the bonding agent will then bond to itself when you bring the two pieces together.

Having now become an expert in tools and glues, you can tackle the job of building model rockets with considerable confidence that *maybe* you can beat Murphy's Law: "If anything can go wrong, it will." Provided you practice what you've learned in this chapter.

4

MODEL ROCKET CONSTRUCTION

Today, most model rockets are built from kits that contain all of the parts and components necessary to complete the model (except glue and paint). As a result, many prefabricated parts are now available, and a model rocketeer isn't required to make highly detailed or difficult parts that call for special techniques or power tools such as a lathe. However, the hobby has always contained a cadre of dedicated model rocketeers who prefer to build their model rocket airframes from scratch. This is especially true when it comes to building scale models, as we'll see in chapter 16. Scratch builders will use whatever parts are available from the manufacturers, but they are forced to make their own parts if these aren't available.

The only model rocket components that aren't made by model rocketeers are model rocket motors. Making your own model rocket motors isn't smart; it's suicidal.

However, it doesn't make any difference whether you're a kit builder or a scratch builder. Certain universal techniques are used by both kinds of modelers.

I'm devoting an entire chapter to this general subject because in my large collection of hobby how-to books many authors have assumed that their readers are experienced and don't need a basic explanation of how and why. I find the *why* to be very important to the technically curious person who becomes involved in model rocketry. So, back to square one.

NOSES

The nose is the front end of the model rocket airframe. After this astounding and illuminating statement, let me add that its shape can vary widely. We used to call it a nose cone. However, since its shape was rarely that of a true cone, we soon started calling it just a nose.

Some of the most common nose shapes are shown in Figure 4-1.

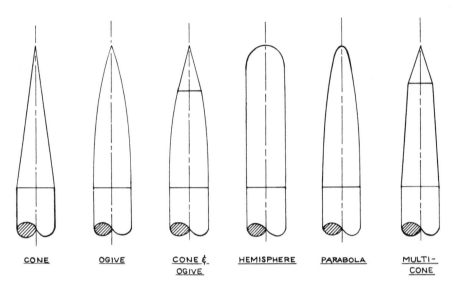

CONE   OGIVE   CONE &   HEMISPHERE   PARABOLA   MULTI-
                 OGIVE   CONE

FIGURE 4-1   Some common model rocket nose shapes.

Unless your model rocket uses a special kind of recovery system such as rearward ejection (see chapter 12), the nose must always be free to slide forward and come off. Therefore, the back end of the nose is cut down to form an internal shoulder or, technically, a tenon, which will slide inside the body tube and hold the nose in place.

The base diameter of the nose should match the *outside* diameter of the body tube. The diameter of the shoulder or tenon should be slightly less than the *inside* diameter of the body tube so that it will slip-fit inside the tube. It's better to have the nose tenon a little loose because you can always build it up or "shim" it with cellophane or paper tape wrapped around the tenon to get exactly the right fit. If you have a nose tenon that is too big for the body tube, you'll have to trim the tenon. If it's a balsa nose, you can "scrunch" it by rolling the tenon along the edge of a table, crushing the balsa until you get a proper fit. With a plastic nose, you'll have to use a small file, sandpaper, or even your modeling knife to cut down the tenon's diameter.

Sometimes after many flights, carbon particles and other debris from the ejection charges build up on the inside of the body tube, making a tight fit for the nose tenon. That's another reason to have a smaller tenon than necessary, one that's shimmed to fit, because you can always remove the shim tape to make the tenon fit into the cruddy body tube.

A very large number of noses are now available from model rocket manufacturers. Noses are usually made from turned balsa wood or from injection-molded or blow-molded polystyrene plastic. *Never use a metal nose!* If you must increase the weight of a nose to achieve the proper balance point and flight stability of your model—as we'll discuss later—fill a hollow plastic nose with the required weight of plasticene modeling clay, or for a wooden nose drill a hole in the rear end and fill it with clay.

Don't stick a metal pin or nail into the tip of the nose to simulate a nose probe antenna. It could turn your model rocket into a rocket-powered dart if something goes wrong in flight. You really don't need to have a sharply pointed nose on your model rocket, anyway. Slightly rounded nose tips have less air drag, as we'll see later.

You can make your own special nose if you have an electric drill in your workshop. Drill a ¼-inch hole in one end of a balsa block. Glue a ¼-inch hardwood dowel into the block so that it protrudes about 1 inch. This gives you something to tighten into the chuck of the electric drill (balsa is too soft to be held well in a chuck). When the glue dries so that the balsa block doesn't separate from the dowel as you start to spin it in the drill, insert the dowel in the chuck, turn on the drill motor, and *carefully* carve the balsa down to the desired nose shape using a file and very coarse sandpaper. This isn't an easy job because the block starts off square and you have to turn it into a cylinder first. Be prepared to create some unusual egg-shaped noses at first. It may take several tries before you learn how to turn a round nose of the desired shape.

The ultimate, of course, is to turn the nose on a precision lathe. You may need to do this when you get into scale model work and find yourself required to make an unusual nose shape and hold dimensional tolerances to a thousandth of an inch.

BODY TUBES

Body tubes for model rockets are usually made from thin-walled paper tubes. Because such tubes are not easy to make, most model rocketeers buy them ready-made from hobby shops or by direct mail from the manufacturers.

Body tubes are available in diameters ranging from 0.197 inch to 6 inches or more (see Figure 4-2). Common lengths are 18 and 24 inches. Buy a body tube in the longest available length and cut it to custom length. You will have some body tube scraps left over, and you'll find plenty of uses for these as motor mounts, stage couplers, spacers, payload supports, etc.

To cut a tube to the required length, first measure the desired length of tube with your steel rule and mark the tube. Then wrap a piece of paper or card stock around the tube at the mark and draw a line around the tube's circumference using the card edge as a guide. Cut the tube with an X-Acto knife. Make several passes around the tube with the blade, cutting only a little on each pass. Don't try to cut all the way through the tube on the first pass or you'll mess up the tube. After some practice, you should be able to cut a body tube so that you can't tell it was cut to custom length.

Bodies for scale models that aren't cylindrical in shape can be made by the hollow log technique. A balsa block is first turned on a lathe or carved to the desired external shape. Once this has been done, the modeler has two choices in completing the model. Sometimes a hole can be drilled down the center of the body block so that a regular paper body tube can be inserted and glued into the block; this tube then holds the motor mount and other internal parts. However, this technique usually results in a heavy model. The true hollow log technique requires that you cut the body block lengthwise in two with a very

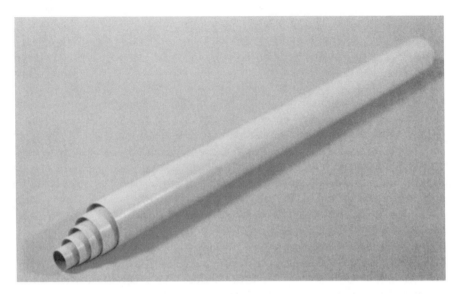

FIGURE 4-2   Model rocket body tubes come in a variety of diameters.

thin saw blade. Hollow out both halves until the sidewalls are about $\frac{1}{16}$ inch thick, just enough so that you can see light through the balsa. Glue a body tube down the middle of one of the halves. Glue the two halves together again. With a little sanding and filling, you won't be able to see where the joint is, and you will have a very lightweight, thin-walled balsa body shell.

MOTOR MOUNTS

A motor mount in a model rocket serves two purposes: (1) It holds the model rocket motor firmly in place so that it can't move forward under thrust or backward and out of the model when the ejection charge activates. (2) It holds the model rocket motor straight and in alignment with the model rocket airframe so that the thrust is directed along the center line of the model.

If the motor is not held firmly in the model, the thrust can ram it forward so that it comes out of the front end of the body tube. If this happens, the motor usually reams the body tube, taking the wadding, recovery device, shock cord, and every other internal part out of the front end with it on a short upward flight. This is not good for your model.

If the motor is not mounted firmly in the model, the puff of gas from the ejection charge can pop the motor backward out of the model. When this happens, the recovery device usually doesn't deploy. The model comes down like a streamlined anvil, usually saving the modeler the trouble of disposing of it because it self-destructs upon meeting that region of very high drag called the ground. This causes what model rocketeers call a DSE (detectable seismic event). However, these "death dives" are *not* funny because they're dangerous. So use a proper motor mount with a motor retaining clip.

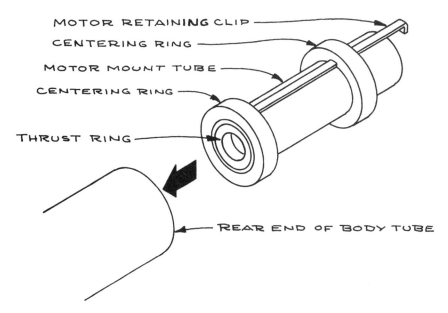

FIGURE 4-3   Drawing of a typical motor mount for positioning and retaining a model rocket motor in a larger body tube. Centering rings and motor mount tube can be eliminated if the body tube is the same size as the motor. But don't eliminate the motor retaining clip.

A motor mount usually consists of several basic parts: (1) the motor mount tube into which the model rocket motor slides with a slip fit; (2) the thrust mount or ring that prevents the motor from ramming forward during thrust; (3) centering rings that will center the motor mount in a larger body tube; and (4) a thin, springlike motor retaining clip (see Figure 4-3).

In simple and inexpensive models, the body tube itself has the proper diameter for the motor to slip into without the need for a motor mount tube and centering rings. A thrust mount ring is usually a small paper doughnut glued into the motor mount tube to prevent the motor from going forward. It has a hole through its center to allow the ejection charge gas to pass. Centering rings are sometimes large paper doughnuts or, for larger body tubes, cardboard or even plywood disks with holes in their centers. Sometimes they're slotted to clear the motor clip.

Nearly all model rocket designs now incorporate a motor clip to retain the motor. If the model design does not have a motor clip, you must be certain that the motor is installed *tightly* in the motor mount tube for each flight. Build up the motor diameter using paper tape. The motor should fit tightly enough that you can't pull it out of the model with your fingers. But you should be able to remove it by giving it a firm, sustained pull with pliers. If the motor casing is flush with the rear end of the motor tube so that you can't grasp it with pliers, you'll have to push the motor out using a $\frac{1}{4}$-inch wooden dowel about 12 or 18 inches long. Just stick the pusher rod dowel down the front end of the model, being very careful that you don't push the entire motor mount out of the body tube in the process of pushing out the motor casing. The difficulties of getting a motor to stay in the model and be easily removable have led to the almost universal acceptance and use of a motor clip today. But there are always those modelers who don't use them.

If you do a good job of assembling the motor mount and gluing it securely into the model, it will never come out. And your model rocket motors will slide in and out easily but not come out in flight.

FINS

A model rocket must have fins or stabilizing surfaces on its rear end in order to fly properly. You should not experiment to determine the truth of this statement. I'll prove it in chapter 9.

The fins on a model rocket are like feathers on an arrow. They keep the model going straight in the air. Model rockets do not fly in the ordinary sense of the word. Their fins are not wings to provide lift forces to keep them aloft. Fins are stabilizing devices to ensure a safe and predictable flight path.

Some beginners' model rocket kits have molded polystyrene plastic tail fin units that fit over or into the rear end of the body tube (see Figure 4-4). They eliminate the most difficult and time-consuming task a beginner faces in building a model rocket: getting the fins on straight and strong. If fins are not attached correctly, the model can fly erratically or not at all. These plastic fin units provide the sort of true and predictable flights required of beginners' models.

With some plastic tail assemblies, it's important to make sure that they won't slip off the rear end of the body tube when the motor thrust accelerates the model off the launch pad and into the air. When this happens—and it occasionally does if a beginner doesn't check it—the model leaves the fin unit behind and becomes an unstable finless rocket. You never want to have it hap-

FIGURE 4-4 Some available commercial plastic tail assemblies with fins molded as part of the tail unit.

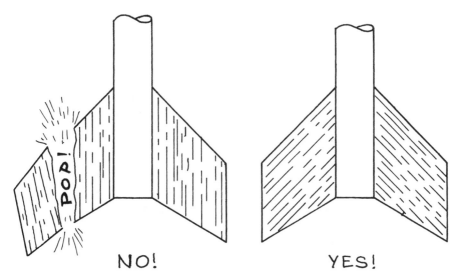

FIGURE 4-5   Always cut fins so that the balsa wood grain follows the leading edge of the fin pattern.

pen again because the model thrashes around in an unstable manner as it leaves the launch rod.

The easiest way to prevent this from happening is to wrap a layer of cellophane tape around the rear end of the body tube, enlarging the tube just enough for the tail assembly to have a snug fit. Some modelers glue on the plastic tail assembly to make sure it doesn't leave the party unannounced. But if a fin breaks on landing, a model rocket with a glued-on plastic tail assembly usually must be trashed because it may not be possible to glue the fin back on, even with cyanoacrylate bonding agent.

However, most model rockets use fins cut out of cardboard or balsa wood that are glued to the rear end of the body tube. Some models have prefabricated molded plastic fins. Most balsa-fin models come with die-cut balsa fins that don't have to be cut out with a modeling knife. But it's a good idea to run around the die-cut lines with a sharp No. 11 X-Acto knife to ensure that the die has cut all the way through the balsa sheet and that the fins will pop out of the sheet without breaking in the process.

If the fin isn't die-cut, you'll have to cut it out of the balsa or cardboard sheet yourself. Most kits with this method of fin construction have a printed fin planform template or the fin outlines printed on the balsa sheet. In the former case, draw a pencil line around the template onto a balsa sheet and cut along the pencil line with a modeling knife. If the fin planforms are printed on the sheet, cut carefully along the lines with your modeling knife.

Important: When laying out a fin pattern on a balsa sheet, align the fin so that the grain of the balsa runs parallel to the leading edge of the fin (see Figure 4-5) or *outward* from the body tube. This is not a cheap trick to get you to use more balsa. If the balsa grain runs parallel to the body tube, the fin will not be strong enough and will break easily. Die-cut fins or printed-on-balsa fin patterns are already oriented with the balsa grain parallel to the fin leading edge. To make sure you never forget this important point, here is a memory jogger to help you: *the grain always runs out!*

When you cut fins from the balsa sheet, try to cut squarely across the balsa sheet with the knife blade. To cut a straight line, cut along the edge of your metal rule. Making a square, straight cut may be difficult to do at first, but keep at it. Each of us develops a personal style of holding the knife and making the cut.

Large fins can be made stronger by using thicker balsa sheets. For most small sport models, a $\frac{1}{16}$-inch balsa sheet is usually strong enough. For larger models or for large fins, use a $\frac{3}{32}$-inch or $\frac{1}{8}$-inch balsa sheets. Fins can also be cut from plastic sheets or from modeling plywood.

To further strengthen a large balsa fin, glue paper on both sides of it. You can also cover balsa fins with model airplane tissue and other coverings. Such a covering makes a very strong fin. It also eliminates balsa wood grain and makes the fin easier to finish smoothly.

Once you've cut out all the fins, stack them together and sand the stack to make sure all the fins are the same size and shape. This may not seem to be important with sport models, but it can become critical with high-performance competition or scale models. So develop this stack sanding habit from the very start in your model rocket activities.

Your model will fly perfectly well if you leave all fin edges square. But performance and altitude can be nearly *doubled* if you take the time to put a streamlined *airfoil* on the fins. Some common ones are shown in Figure 4-6. The simplest airfoil merely has the edges rounded except at the fin root where you glue the fin to the body tube. You can round fin edges quickly and easily with your sanding block.

In the early days of model rocketry, we put sharp-nose wedge-shape airfoils or double-convex airfoils on fins, emulating the appearance of the big rockets at the Cape. But the big ones are designed to fly at supersonic and hypersonic speeds, whereas model rockets rarely attain half the speed of sound. Therefore, the best airfoil for a model rocket fin is one with a rounded leading edge and a tapered trailing edge as shown. It takes a little time to do this shaping with your sanding block. You may mess up a couple of fins in the process. But cut out some more fins and try again. The effort will pay off handsomely in improved performance. Don't round or taper the fin edge that will be glued to the body tube! (Yes, I've seen rocketeers do it!)

Some modern model rocket kits come with molded plastic fins that have the planforms and airfoils already formed. Thus, you don't have to worry about these factors. How many fins should you put on a model rocket? The bare minimum is three fins in the triform arrangement shown in Figure 4-7. Some models have a four-finned or cruciform arrangement. Models are also made with five or six fins. However, using more than four fins usually increases the air drag and reduces performance.

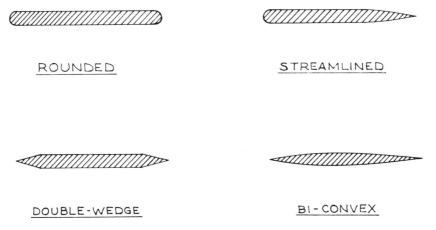

FIGURE 4-6 Cross-section drawing of some basic model rocket fin airfoils.

Always put the fins at or near the rear end of the body tube. The reasons for this and the reasons that fins have certain sizes and shapes are discussed in detail in chapter 9.

You should *never* put fins near the nose of the model. Nor should you put fins anywhere except at the rear end unless the kit instructions tell you to do so and tell you precisely where to put them. Stability begins to get very tricky when you put the fins up front. You could end up with a model rocket that

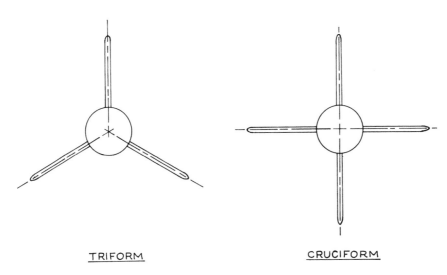

FIGURE 4-7 Drawing of the rear of two model rockets showing three- and four-fin configurations.

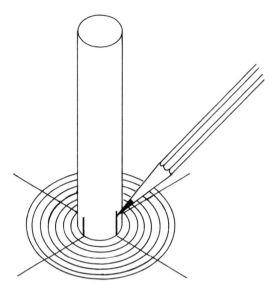

FIGURE 4-8   How to use the fin placement
guide shown in Figure 3-12.

wants to fly backward and can't. This will provide far more excitement than
you want, believe me.

You can locate the fins on the body tube by using the fin placement guide
shown in Figure 4-8. Carefully attach the fins so that they stick straight out
from the body tube and are perfectly aligned with the model. If the fins are
canted or crooked, the model may fly at an odd angle, spin, fly erratically, or
otherwise behave in an unpredictable and dangerous fashion. If you put on the
fins straight and true, the model will fly straight and true. Don't forget that the
fins are the model rocket's stabilizing system.

For most small model rockets using motors up to Type D, simply gluing
the fins to the outside of the body tube will provide enough strength to keep
them on the model during flight. Be sure to use a double-glue joint. If you do,
the fins will break and the body tube will tear before the double-glue joint
cracks and fails.

For larger model rockets, advanced fin attachment methods have been
devised and must be used. We'll cover these in chapter 14 on large model rockets.

SHOCK CORD MOUNTS

The shock cord is the line that attaches the nose and recovery device to the
body tube. The term was first used by Orville H. Carlisle, the inventor of
the model rocket, because early shock cords were short and had to absorb the
shock of the nose flying off the front of the model and being stopped before it
slowed down.

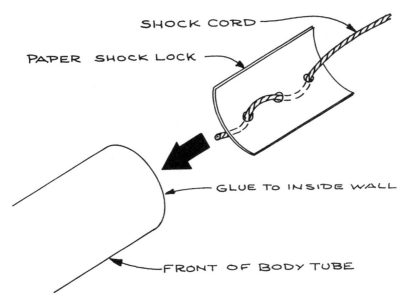

SHOCK CORD

PAPER SHOCK LOCK

GLUE TO INSIDE WALL

FRONT OF BODY TUBE

FIGURE 4-9   The paper shock lock developed by the author for attaching the shock cord to the inner surface of the front end of the body tube. Be sure that the shock lock is glued back from the front edge of the tube by an inch or more so that it does not bind the shoulder of the nose.

A shock cord keeps the nose, the recovery device, and the rest of the model rocket together during recovery so that you don't have to recover many pieces separately. It's usually difficult enough to recover the whole model tied together. However, some model rocket designs separate into two or more components during recovery, each part equipped with its own recovery device.

A shock cord can also be considered to be a bungee just like the elastic cords used by bungee jumpers and parachutists. The first shock cords were $\frac{1}{8}$-inch-wide rubber strands identical to those used in rubber-powered model airplanes. This type of shock cord is still used in many model rocket kits. However, it has been replaced by the more robust cloth-and-rubber elastic cord and even by stout cotton twine. The advent of heavier model rockets has also required the use of stouter shock cords.

A shock cord doesn't have to be elastic if your model is light and you use enough shock cord. An 18-inch or 24-inch length of stout cotton kite twine will work. Shorter lengths of twine will break. A twine shock cord will last for dozens of flights. A ball of cotton kite twine purchased in a hobby shop will provide you with shock cords for a couple of years. Use only cotton twine; it won't be melted by the heat from the ejection charge. Twine made from nylon or other artificial fibers will melt (exception: Kevlar, as we'll see in a moment).

The classical methods of attaching the shock cord to the body tube have been to (1) slit the body tube, feed the cord through the slit into the body tube,

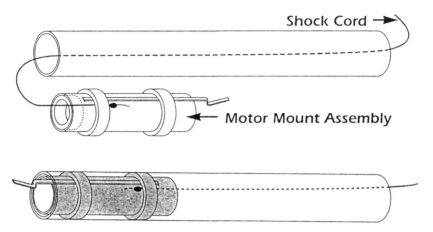

FIGURE 4-10   In Quest model rocket kits, a shock cord made of high-temperature Kevlar is attached to the motor mount assembly and a length of elastic is tied to the forward end of this, then to the base of the nose. (Drawing courtesy Quest Aerospace, Inc.)

and glue the short end down flat on the outside surface of the model, or (2) glue it to the inside front surface of the body tube with a shock mount. In 1969, I invented one of the first internal shock mounts, shown in Figure 4-9. Other variations have been used by various model rocket kit manufacturers.

Bill Stine at Quest Aerospace, Inc., developed a new method in 1991. This is shown in Figure 4-10. A length of Kevlar twine is attached to the front of the motor mount. Tied to the front end of the Kevlar is elastic cord that is in turn tied to the base of the nose. The Kevlar will withstand the high-temperature puff of gas from the model rocket motor's ejection charge. This system allows the shock cord to be firmly attached to the thrust structure of the model rocket without marring the outside of the body tube or even partly restricting the inside of the tube.

The new shock cord developments have reduced in-flight separation of the model at recovery ejection. The new shock cords will probably outlast the models in which they're installed.

RECOVERY DEVICES

If you've spent a lot of time building a good model rocket, you'll want to get it back after its first flight in a condition to fly many more times. In addition, even a 2-ounce (57-gram) model rocket falling freely out of the sky in a streamlined condition isn't safe. Those are two good reasons why you should always use a recovery device. Making and installing a workable, reliable recovery device is simple and easy, and model rocket motors are designed to expel such a device from the model in flight.

All the model rocket motor types used in single-stage models or in the upper stages of multistage models are made so that they'll produce a retro-thrust puff of gas at a predetermined time after ignition of the motor. This puff of gas is used to activate or deploy the recovery device. Chapter 5 explains in detail how this is done. Briefly, the motor puffs a bit of gas that pressurizes the inside of the body tube, pushing the wadding and recovery package forward, dislodging the nose, and permitting the recovery package to exit from the model and deploy in the air.

Many types of recovery devices have been developed and used by model rocketeers. Some of the successful ones are described in detail in chapter 12. For now, we'll take a quick overview of the two most commonly used ones.

The simplest recovery device is a crepe paper or plastic streamer that's tied to the base of the nose cone. When ejected from the model, it streams in the wind and flutters, slowing the model's descent. Streamers are used on small models weighing less than 3 ounces (85 grams). They are also used on small high-performance models that ascend to high altitudes or on models that are flown from small fields in windy conditions. A model with streamer recovery will return to the ground more rapidly than one with a parachute, and it will not drift as far in the wind.

A streamer is a long, narrow, rectangular strip of crepe paper or thin plastic film. It's usually 1 to 2 inches wide and 12 to 24 inches long. A fairly standard streamer is 1 inch by 12 inches. The best length to width ratio is 10 to 1—its length is 10 times its width. The streamer should be of brightly col-ored material, preferably orange or red, so that it can be seen against the sky, on the ground, or in a tree.

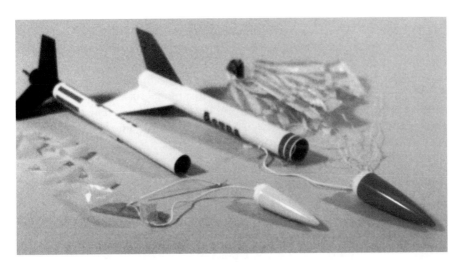

FIGURE 4-11   Two model rockets: one with a streamer for recovery and the other with a plastic parachute for a slower descent.

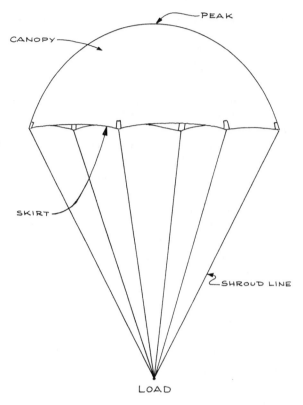

FIGURE 4-12   Drawing of the parts of a typical parachute.

The most obvious recovery device is a parachute. Most model rocket parachutes are made of polyethylene film about $\frac{1}{1,000}$ inch thick, although some kits have silk or nylon parachutes. Commercially made model rocket parachutes are printed or decorated in bright and contrasting colors and patterns so that they can be seen more easily. Most model rocket parachutes have six or eight sides with six or eight shroud lines, respectively. The shroud lines are lengths of carpet thread or nylon twine. The shroud line length should be at least equal to the diameter of the parachute or longer if possible. The shroud lines are attached to the parachute by means of tape strips.

Be sure to attach the shroud lines firmly to the edge of the parachute canopy. When a parachute is ejected from a model rocket and opens in flight, it sometimes snaps full of air with a loud pop that can be heard on the ground more than 100 feet below. This action puts a tremendous strain on the parachute and its shroud lines. If one or more of the shroud lines pulls off, the parachute will lower the model more quickly.

The larger the parachute, the farther the model will drift in the wind. If you put a big parachute into a little model rocket powered by a high-power model rocket motor, you'll probably never get it back. It will drift for miles before it lands.

A parachute 12 inches in diameter is usually adequate for models weighing up to 3 ounces (85 grams). Models with 18-inch parachutes have set world duration records. I've lost count of the number of model rockets I've lost under 18-inch parachutes. At the First International Model Rocket Competition in Dubnica-nad-Vahom, Czechoslovakia, in 1966, my parachute duration competition model with an 18-inch parachute was found 17 miles away by a Czech forest ranger who was somewhat astounded to discover a small rocket with American markings in the middle of the Little Tatra Mountains of Europe.

PAINTING

Unless your model rocket kit comes precolored, paint it. If you do a good job, it will look better. After all, you are proud of it, aren't you? If it's a model with balsa fins, it will perform better if it's painted.

Model airplane "dope" isn't recommended for painting model rockets. It's designed to shrink as it dries. When used on model rockets, this feature can cause the fins to warp. Model airplane dope also melts or crazes the surfaces of polystyrene plastic parts.

Hobby and craft stores carry many types of paint that can be used on model rockets. Most of these are *enamels* or *acrylics*. Testors is one of the many companies that make model and craft paints.

Enamels are usually too thick, leave a lumpy finish, and change color with age (white becomes yellowish). Most model rockets that are brush painted with enamels by beginners look pretty bad. Sometimes I believe the modeler smeared on the enamel with a cotton swab. Acrylic paints can be thinned with water, go on smoothly without showing brush strokes, do not change color with age, dry to the touch in 30 to 60 minutes, produce a water-resistant surface, and don't smell.

I rarely paint a model rocket with a brush. I use spray paint. And I don't use spray paint from the little cans sold in the hobby shops. I go to a hardware or paint store and buy a big can of spray paint. The best is Krylon. The cheaper brands can run and spoil the paint job. To paint your model rocket using these spray paints, put the model on a paper wand and give it several thin coats of paint for the best result. Don't try to get a complete, all-covering paint job with one coat. The paint will usually run. "Dust" the paint on the model. You can put on several light coats and have them dry in less time than it takes for one thick, globby, runny coat to dry.

Airbrushes (small paint sprayers) are now available at reasonable prices in hobby shops. These are ideal for painting model rockets. I have several of them and have actually worn out two of them over the years. Water-based acrylic paints are best to use with an airbrush because you can thin the paint with water and clean out the airbrush easily with water when you change paint colors or get ready to put things away. Pressure for an airbrush usually comes from a can of pressurized carbon dioxide sold for this purpose in hobby shops.

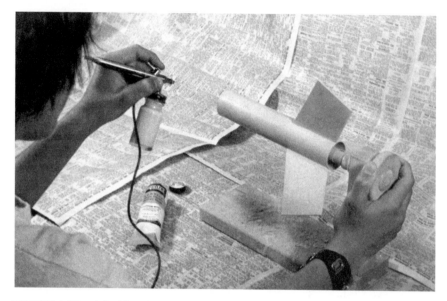

FIGURE 4-13   A hobby airbrush provides a beautiful, even coat of paint and is well worth the expense. Be sure to use lots of newspaper to prevent the overspray from getting everywhere.

These are okay for quickie jobs, but you'll soon spend a lot of money buying pressure cans. So get a small air compressor for your airbrush. It will be a little costly but will quickly pay for itself if you build a lot of model rockets and do a lot of spray painting. Some airbrushes will work on the output of a tank-type household vacuum cleaner; attach the hose to the end that blows, not sucks!

Paint your model in bright colors that can be easily seen in the air and on the ground. Good colors are fluorescent orange and red-orange as suggested earlier. A model rocket is most highly visible when painted one solid color with perhaps one black fin. Painting it in many colors will tend to camouflage it and make it hard to see.

Fluorescent colors are available in hobby store spray cans or in bigger Krylon spray cans. Fluorescent colors *must* be sprayed on a model rocket because brush painting leaves a streaky coat. Fluorescent colors must be applied over a base coat of flat white.

The best nonfluorescent colors for maximum visibility depend on the general sky color and condition in your locale. Against the clear, blue skies of the American West, the best colors are white, orange, or yellow. Against the cloudy, gray, hazy skies of the Midwest, the South, or the East Coast, dark colors such as maroon or black work well. Silver takes the reflected color of the sky, and silver-painted model rockets tend to disappear against the sky. Greens, browns, and sky-blues blend too well with the surroundings. One not-too-bright model rocketeer once painted his rocket in accurate military camouflage colors that really worked; he couldn't find his model after it landed on the ground.

DECALS AND TRIM

Most competition model rockets don't carry much decoration other than the contestant's sporting license number to identify it as required by national rules. Competition model rocketeers have learned that putting decals and strongly contrasting paint patterns on a model rocket tends to make it more difficult for the human eye to see and follow. This is important because people on altitude trackers or timing the flight with stop watches *must* be able to see the model. Note that military camouflage and the natural coats of animals use a number of contrasting colors in random patterns; this makes tanks and animals hard to see. The same principle applies to model rockets.

However, the application of decals and trim to a sport model results in a striking change in its appearance. The smooth, streamlined, uncomplicated shapes and lines of the model suddenly seem to come alive. The model begins to look like its big brothers at the Cape. To obtain a good-looking model that resembles a real rocket, it's important to know what kinds of markings and trim to use. This is because the color, patterns, and markings of the real ones have definite functions.

The most important marking is the roll pattern or regular stripes, checks, or other marks around the body. At the Cape, a roll pattern is painted on a rocket for photographic data purposes so that engineers can determine from camera films how the rocket rolled or tilted in flight. If you roll one of your models in your hand, you'll see how the roll pattern marking changes appearance and how it's possible to determine roll rate from that change in appearance.

Usually, one fin is painted black or some contrasting color. This is to provide a roll reference point. Fins sometimes have different paint patterns to identify them; engineers looking at the camera films of the flight can tell which side of the rocket they're seeing.

Most NASA space vehicles and sounding rockets carry a painting of the American flag and the words *United States* or *USA* on the body of the first stage. The star-and-bar military insignia is never applied to a nonmilitary NASA rocket. That insignia, however, and the words *United States Air Force* or *USAF* usually are placed somewhere on the body of an Air Force rocket. Often, an Air Force rocket has some sort of squadron or command emblem or other marking on it.

Numbers are usually applied to fins but may also appear on the body. Numbers and letters are usually applied in an orientation that permits them to be read when the rocket is in the launching or flight attitude.

Follow the instructions with the kit or the decals in order to apply the decals properly. Other markings can be applied to a model using an India ink pen.

Today, there is a plethora of powerful graphics software available that allows you to design you own model rocket graphics and print them out on an ink-jet printer. These graphics can then be made into water slide decals using a commercially available product like the SuperCal decal system. The Super-Cal system includes a special decal paper that you use in your color ink-jet printer. After you print the sheet, you spray it with decal coating, let it dry, and you've got a high-quality set of water slide decals to apply to your rocket!

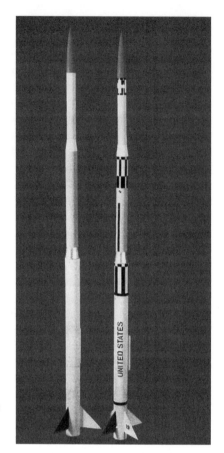

FIGURE 4-14   The proper placement of a few decals on a model rocket will greatly improve its appearance.

After all decals and markings have been applied and are dry, apply one light coat of dull or gloss transparent spray over the entire model, depending on whether you want it to have a flat or glossy finish. Dulling spray will kill the shine of decals and will also make the model look bigger.

STORAGE AND CARRYING

Although your model rocket may be strongly constructed and capable of flying through the air at half the speed of sound, it can be easily broken when it's on the ground. Little Brother may get to it. Fido may decide it's a superior and very tasty new kind of bone. Big Clod may step on it. Fumblefingers may drop it 3 feet to a concrete floor where it will shatter into 16 jillion little pieces, most of which you'll never be able to find. Or it may simply get trashed by an overzealous mom who doesn't understand that you're trying to get a hundred flights on that ratty old model rocket.

Put your models away between flight sessions. Make a display rack for them. Or hang them from the ceiling by threads. Put them in a big drawer—but be careful not to shut the drawer on the fins. A big cardboard file box is wonderful for storing model rockets.

To keep a model rocket from being damaged in storage and in transit to the flying field, put it in a plastic bag. The air trapped inside the bag acts as a cushion to protect the model. I've carried model rockets on airliners in plastic bags inside a cardboard box and had them arrive without the slightest damage or paint scratch. (Some of my model rockets have thus been as high as 37,000 feet or more!) I always carefully open the plastic bag in which the model rocket kit comes because I save that bag and put the completed model in it.

Now that we've covered some of the basics of model rocketry, we can begin to talk about the more complex and advanced parts of the hobby. Before you go on, however, read these first four chapters again just to make sure you understand what we've been talking about and don't forget it. Even after you've read the rest of the book, come back and review these fundamentals from time to time. That way, you'll have a sound understanding of them and won't be held back by making basic mistakes all over again.

5

MODEL ROCKET MOTORS

The device that makes model rocketry a hobby and a sport rather than a disaster is the model rocket motor. Some people call it a model rocket engine, but there is a subtle distinction between a motor and an engine. A motor is defined as something that imparts motion. An engine is defined as a machine that converts energy into mechanical motion. Thus, a steam engine or an internal combustion engine is definitely an engine. A model rocket motor is truly a motor. Technically, it is a small reaction device for converting the energy of high-temperature, high-pressure gas into motive power without the use of gears, cams, linkages, pistons, turbines, etc.

There are many kinds of rocket motors. They're usually categorized by the type of propellant they use—liquid propellant or solid propellant—or by some unique form of energy that they convert into motion—nuclear rockets, ion rockets, nuclear pulse rockets, etc.

All model rocket motors developed for use in the hobby of model rocketry thus far are solid-propellant rocket motors. As we will discover, there are several reasons for this. A solid-propellant model rocket motor is an inexpensive, highly reliable package of power that has been designed for use by nonprofessional consumers. It will provide both the propulsive force to thrust a model rocket hundreds or thousands of feet into the air and the means for ejecting the recovery device. It requires no mixing of dangerous chemicals. It may be a preloaded single-use expendable unit or a reusable unit that can be reloaded with prefabricated propellant and other modules designed by the manufacturer for this purpose.

Model rocket motors are the world's most reliable rocket motors. By 2003, nearly 600 million of them had been manufactured and used in the United States alone without producing any severe fire hazard or personal injuries more serious than an occasional burned finger. The few accidents that have occurred have been caused by *failure to read and follow instructions* (does that sound familiar by now?) or by deliberate product misuse.

A model rocket motor appears to be a very simple device. Actually, it's very complicated. Making one costs a large amount of money and requires extensive and very special equipment, plus a lot of knowledge. Strict safety precautions must be used in making model rocket motors. This explains why model rocketeers leave the making of rocket motors to the model rocket manufacturers.

There is no safe way to make a rocket motor of any type. This is a statement of fact, not a matter of opinion. There is also no way that you can make a rocket motor that is as inexpensive and as reliable as a commercial one. A model rocket motor—which now includes a reloadable model rocket motor and its prefabricated reloading kit—is a factory-made device that is subject to rigid quality standards, quality controls, and statistical batch sampling and testing procedures. It's very reliable and will do exactly what it's designed to do. You, as a model rocketeer, don't have to handle dangerous chemicals, worry about whether the motor will have proper thrust, or take extensive and expensive safety precautions.

A model rocket motor is not a toy. You must understand this right from the start. Safe as it has proven to be, a model rocket motor can hurt you if you don't use it correctly and in accordance with instructions.

A model rocket motor is packaged reaction propulsion power for models and should not be used for other purposes. Model rocket motors should be your introduction to the fact that technology will work for you if you handle it properly—and that it can bite you if you don't. Cro-Magnon people learned about fire the hard way when they got burned by their cave fires. Fortunately, you don't have to learn everything in that difficult school of hard knocks called experience. However, you must be willing to listen to people who know something about it and follow their instructions and teachings. Soon you'll know more than your teacher. Harry listened and learned from the experts—people who had worked for Dr. Robert H. Goddard as well as German experts such as Dr. Werner von Braun. Because I listened and learned from my father, I still have all my fingers, my eyes, and the hair on my head after more than twenty-five years of working with rockets of all sizes and types. The conquest of space has few slots for blinded rocket engineers or space pilots without fingers or hands.

As of this writing, there are hundreds of different types and makes of model rocket motors available in the United States that have been tested and certified by the Standards and Testing Committee of the National Association of Rocketry (NAR) as meeting the reliability and performance standards of the NAR and the National Fire Protection Association (NFPA). This is the largest selection available in any country in the world. From this broad selection, you should be able to choose a model rocket motor that meets your requirements. Go to www.thrustcurve.org to find motor performance data online. This site collects and organizes data on certified motors for rocketry hobbyists. You can search for motor data by manufacturer, impulse, or size, then import the data into one of the many design/flight simulation computer programs.

FIGURE 5-1   Some of the model rocket motors manufactured in the United States at the time of this writing. Units are representative of some of the different sizes available.

Most model rocket motors used today are preloaded, single-use, expendable units. We'll use the preloaded expendable model rocket motor as a starting point for learning about motors. Later, we'll discuss reloadable, reusable model rocket motors.

A typical solid-propellant model rocket motor is shown in Figure 5-3 as though it were cut down the middle to expose its innards. (Don't do it! It's safer to look at the drawing!)

The motor *casing* is made from tightly wound paper with carefully controlled dimensions or from special composite plastic. Most model rocket motors use paper because it's strong, fire resistant, and doesn't conduct heat easily. Advanced high-performance model rocket motors usually use composite plastic casings for even greater strength and heat resistance. The dimensions of model rocket motor casings vary according to power, manufacturer, etc. However, there are some basic sizes, as shown in Table 5.

The *nozzle* is made from ceramic formed into a carefully designed size and shape. Advanced model rocket motors may use a nozzle molded or turned from high-temperature phenolic plastic or carbon. You should never alter the nozzle because the slightest change in dimension of the nozzle throat—as little as $1/1{,}000$ inch—could drastically alter the operation of the motor.

The *solid propellant* is a piece of combustible chemical material with controlled burning characteristics. Model rocket motors that have been in use since the hobby started in 1957 normally use *blackpowder,* a classic rocket

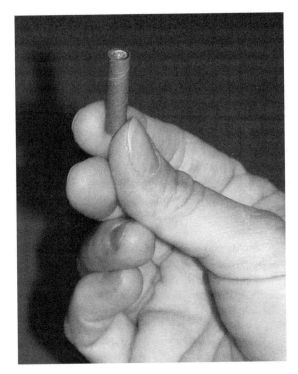

FIGURE 5-2   Quest Aerospace, Inc., has developed the MicroMaxx motor. A 0.25-inch diameter and impulse of 0.32 newton make these tiny motors perfect for flying rockets in small areas.

propellant first used by the Chinese in A.D. 1232 at the Battle of Kai-Feng-Fu. We know a great deal about blackpowder. The most important thing for you to know about it is to leave it alone. Don't tamper with the model rocket motor, and don't try to remove the blackpowder propellant. Since 1980, *composite* solid propellants have been used in model rocket motors. Composite propellants were developed for ballistic missiles and are used in the solid rocket boosters of the NASA space shuttle. The most commonly used composite solid propellant in model rocket motors is a mixture of ammonium perchlorate, synthetic rubber, and other exotic Space Age chemicals to control burning rate, stability, and other factors. Composite propellants have about three times the power per unit weight of blackpowder, but they are more expensive.

Once ignited, the kind of solid-propellant motor shown in Figure 5-3 burns forward from where it is ignited at the nozzle, producing more than 2,000 times its solid volume in hot gas. This gas shoots out the nozzle and produces thrust in accordance with Newton's Third Law of Motion. More about this later.

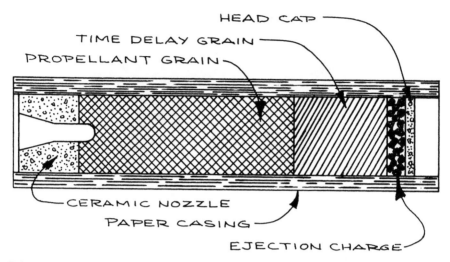

FIGURE 5-3   Cross-section diagram of a typical solid-propellant model rocket motor.

Ahead of the solid-propellant charge is the *time delay charge*. This is a piece of very slow-burning solid propellant that produces very little gas and therefore practically zero thrust. Its action allows the model rocket time to coast upward on its momentum to apogee. If the time delay were not in the model rocket motor, the ejection charge (see below) would deploy the recovery device from the model at a low altitude and a high speed. The model and the recovery device aren't strong enough to withstand this sort of flight behavior even once.

The time delay charge lasts for several seconds, depending on the type of motor. The end of burning of the time delay charge automatically activates the *ejection charge*. This produces a quick puff of gas that pressurizes the inside of the model rocket body tube, pushes the recovery wadding and package forward, dislodges the nose, and expels the recovery device from the model. The

TABLE 5
Model Rocket Motor Sizes

Family/Series	Manufacturers	Diameter × Length	
		Millimeters	Inches
Micro	Quest	9 × 25.4	0.25 × 1.00
Mini	Estes	13 × 45	0.50 × 1.75
Standard	Estes, Quest	18 × 70	0.69 × 2.75
24 millimeter	Estes, AeroTech	24 × 70	0.945 × 2.75
29 millimeter	AeroTech Type F	29 × 98	1.13 × 3.88
29 millimeter	AeroTech Type G	29 × 124	1.13 × 4.88

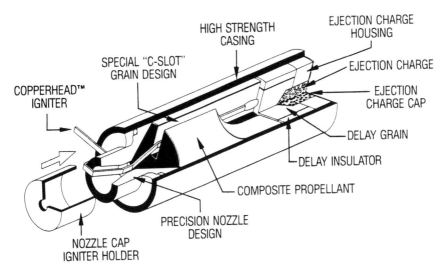

FIGURE 5-4   Cutaway drawing of a typical expendable composite-propellant model rocket motor with igniter installed. (Courtesy AeroTech Consumer Aerospace.)

ejection charge is held in place in the motor with a *head cap* that is either a paper cap or a thin ceramic plug that's shattered when the ejection charge is activated. And that's all there is to it. And it works.

Expendable model rocket motors are good for only one flight. They cannot and should not be reloaded with anything. The casing of an expendable model rocket motor has been designed for only a single use. If you want to cut apart a *used*, expended motor casing (never a loaded one), you'll see that the inner surface of the casing has been charred and ablated during motor operation in the same manner that the old space capsules' heat shields were ablated by atmospheric frictional heat when they returned to earth. This casing ablation weakens the casing to the point that it isn't safe to use it again.

Since 1985, *reloadable* model rocket motors have been developed. A reloadable motor has a metal or plastic casing that is designed to be reused. Reloadable rocket motor technology has been around for a long time. In 1947, the British Jetex rocket motor for model airplanes used a reloadable casing, but its propellant wasn't energetic enough to produce the thrust necessary for model rocket flight. Reloadable model rocket technology grew out of the development of a composite solid-propellant rocket motor to pull a parachute out of an ultralight airplane or a Cessna 150 training airplane in trouble. The technology was developed to a level of safety that permitted it to be used in a manned airplane. Getting it accepted for use in model rocketry required years of additional testing. Reloadable model rocket motor technology is now as safe as classical expendable model rocket motor technology, but using it is more complicated than simply inserting the replaceable, expendable

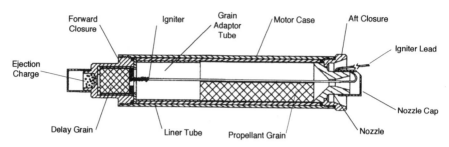

FIGURE 5-5   Cross-section drawing of a typical reloadable model rocket motor. This shows a single-grain module installed. (Courtesy AeroTech Consumer Aerospace.)

model rocket motor. Manufacturers, the NFPA, and the NAR recommend that reloadable model rocket motors be sold and used only by those 18 years of age or older because of the greater complexity of these motors. Current regulations of the Consumer Product Safety Commission also require this.

A typical reloadable model rocket motor is shown in Figure 5-5. The casing, *aft closure,* and *forward closure* are designed to be used again and again. The *motor reloading kit* contains one or more composite-propellant *modules* (reloadables use *only* composite solid propellants), each module encased in a cardboard tube. Also in the reloading kit is a cardboard or plastic *insulator* that is inserted into the empty casing first. The insulator keeps the burning composite propellant from raising the external temperature of the casing beyond 200 degrees C, the same external temperature limit imposed on expendable model rocket motors. (Thus, reloadables can be used in the same kind of models as expendable motors.) Once the propellant is installed, one or more *O-rings* are inserted to seal the aft closure, and the *aft closure* is attached. Then the prepackaged *delay module* and *ejection module* are inserted into the forward end of the casing, the O-rings installed, and the *forward closure* attached. The motor is installed in the model in the same manner as an expendable motor, and ignition of a reloadable is similar to that of an expendable motor.

Because the design and construction of a reloadable model rocket motor is more complex than that of an expendable motor, and because this technology is moving forward and changing rapidly, *be sure to read and follow all instructions of the manufacturer. Do not load or reload a reloadable model rocket motor with any material not provided by the motor manufacturer in the reloading kit.*

The primary reason for the existence of a reloadable model rocket motor is low cost. Most reloadables are used in the upper end of the model rocket motor power ranges. Large expendable model rocket motors get to be expensive. However, reloadables have about one-third the cost per flight of comparable expendables. But a reloadable motor has a high up-front cost that offsets the low per-flight cost unless you are doing a lot of flying. It takes between four and six flights of a reloadable to achieve any savings over using an expendable.

It's also messier and it takes time to clean up the reusable casing and components after a flight. However, some model rocketeers enjoy this additional involvement and activity. Again, *always follow all the instructions that come with any model rocket motor, expendable or reloadable.*

Expendable or reloadable, a lot of information is printed on the model rocket motor casing and even more in the package and instructions that come with the motor. *Read these!* And check for the statement "NAR Certified" on the package or the motor casing, meaning that the Standards and Testing Committee of the NAR, a nonprofit representative for consumer rocketry in the United States, has tested samples of that type and make of motor and has determined from these tests that the motor type meets or exceeds the strict set of performance and reliability standards developed by the NAR and the NFPA. If a model rocket motor or its package doesn't have "NAR Certified" printed on it, you should be wary of it. In most states, it's against the law to sell or use a model rocket motor that doesn't have NAR certification.

All NAR-certified model rocket motors carry on their casings the universal U.S. model rocket motor code that tells what kind of motor it is and how it will perform. This NAR motor code is simple. It consists of a letter, a number, a dash, and a final number. A typical example might be B6-4. The *first letter* of the code indicates the *power range* of the motor. How do you figure the power of a model rocket motor? In terms of *total impulse,* which is a factor derived by multiplying the average thrust by the thrust duration. Or, more accurately, total impulse is the area under the motor's thrust-time curve.

Sound confusing? Well, it won't be if you take it one step at a time. The *thrust* of a rocket motor is the amount of force, or push, produced when it's operating. The jet of gas rushing out the nozzle at supersonic speed produces a force according to Newton's Third Law of Motion. Stated simply, for every acting force, there is an equal and oppositely directed reacting force. Written as an equation, the universal shorthand of science and technology, this is $MA = ma.$

The thrust of a model rocket motor is rarely constant. It changes with time. Therefore, we must further define which thrust level we are speaking about. There are several. *Maximum thrust* is the highest amount of force produced by the motor during operation, regardless of when that occurs during the period of operation. *Average thrust* will be defined later.

Because model rocketry is an international sport, motor performance specifications are in the units of the international metric system—more specifically, in terms of the meter-kilogram-second (MKS) system. Motor thrust is therefore given in terms of *newtons,* named after the aforementioned Sir Isaac Newton. A newton is defined as the force required to accelerate 1 kilogram (2.2 pounds) of mass at a rate of 1 meter per second per second (3.28 feet per second per second). It's easy to convert newtons of force to pounds of force:

1 pound of force = 4.45 newtons

The model rocket motor begins to produce thrust at the instant *ignition* of the propellant occurs. *Burnout* is the instant that the motor ceases to pro-

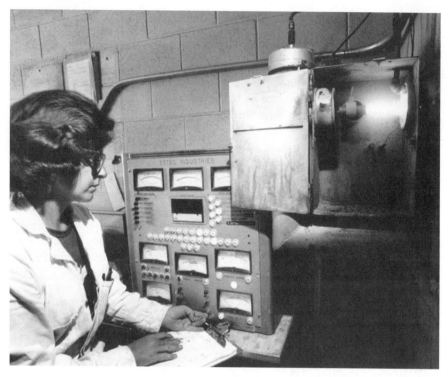

FIGURE 5-6   Static testing a model rocket motor to determine its performance. Manufacturers have developed complex electronic testing equipment to ensure high product reliability. (Estes Industries, Inc. photo.)

duce measurable thrust. The length of time that thrust is produced is called *duration*, and it is measured in seconds of time. A duration of 2 seconds is very long for a model rocket motor. The time interval from ignition to maximum thrust is known as *T-max*. This is an important parameter to know if your model rocket is heavy; it will help you choose the proper length of launch rod to use, as we'll see later.

Please don't be put off by the continual references to things we'll discuss later. We have to take this one step at a time.

To accurately determine thrust, maximum thrust, duration, T-max, and other performance characteristics, a model rocket motor must be given a static test. It is fastened into a fixture on the ground with measuring and recording devices attached. Then it is ignited and operated.

The data record of a static test produces a thrust-time curve such as the generalized one shown in Figure 5-7. This is a typical thrust-time curve for most model rocket motors. Notice how the thrust rises rapidly after ignition to maximum thrust. This accelerates the model rapidly to high speed, ensuring that it has sufficient airspeed for the fins to stabilize it when it leaves the

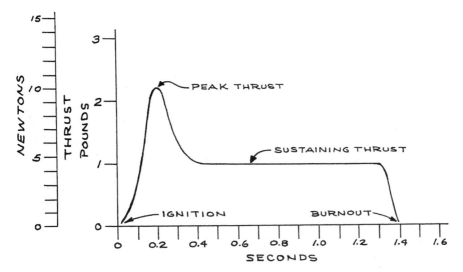

FIGURE 5-7   Typical thrust-time curve of an end-burning solid-propellant model rocket motor showing various events.

launch rod. Then the thrust settles back to a lower value, a sustaining thrust that accelerates the model to higher speeds during the climb.

When a model rocket motor is static tested, it is no longer usable for flight (unless it's a reloadable, but even reloadables must be prepared for another operation with a reloading kit). How do we know that other motors of this type are going to produce the same performance? Answer: We don't know for certain, but we can infer that they will. This inference is based on confidence gained from testing a large number of motors of the same type as well as on statistical sampling and proven testing techniques. You can delve into this area of mathematics in detail if you're a math shark—and many model rocketeers are. Model rocket motors follow the laws of statistical difference just like everything else in the universe.

Each model rocket manufacturer has a high-precision model rocket motor static test stand. Today, they're a far cry from the simple equipment using springs and rotating drums that were common in the early days. Solid-state electronics, microchips, and computers have made static testing easier and have brought the cost and complexity of a static test stand within the means and capabilities of a high school model rocket club.

Manufacturers subject a random sample—usually a minimum of 2 per-cent—of each model rocket motor production batch to static tests. These check the performance and conformance to NAR standards. If the test sample fails to meet the standards, the entire production lot of perhaps hundreds of motors must either be rejected and destroyed or corrected so that another test sample performs properly.

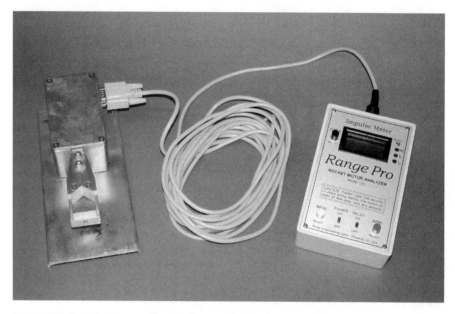

FIGURE 5-8   The Range Pro static test device is an economical test instrument used by schools and rocketeers to measure the performance characteristics of rocket motors.

Over a period of time, a manufacturer gets much information from thousands of static tests. The manufacturer can then have a high degree of confidence that a random sample of a production lot represents the performance of the rest of the lot—that is, the other nontested motors will perform the same as the test sample.

Statistical analysis is a fascinating area of science and mathematics. Few people know of it. But statistical quality sampling tests of products is carried out with products ranging from ballpoint pens and semiconductor computer chips to canned goods and automobiles.

After all, if 600 million model rocket motors have been made since 1957, over 10 million model rocket motor static tests have been done. That's a lot of data! It provides a very large statistical universe. Even a small manufacturer that produces 100,000 model rocket motors a year accumulates data on 2,000 static tests every year.

In addition to the manufacturers' static testing and the testing that is carried out by the NAR Standards and Testing Committee to certify a new motor type, the NAR obtains and tests random samples of model rocket motors from many sources. Committee members buy model rocket motors from hobby shops all over the country and static test them to make sure that the motors available to consumers have the same characteristics as the original test motors submitted to the NAR for certification. And all certified model rocket motor types are tested for *recertification* by the NAR every three years.

Yes, some motors from some manufacturers haven't passed the certification tests and have been denied certification. Some motor types have had their certifications withdrawn for failure to maintain the original performance standards. The NAR doesn't publicize the failures but gives the manufacturer a chance to correct the faults of the motor type. If the manufacturer doesn't or won't, then the NAR notifies various hobby trade organizations and enforcement groups such as the Fire Marshals Association of North America.

For more than half of a century, this philosophy of testing and certification has worked. The NAR believes that model rocket safety is best ensured at the source: the tight quality control of the manufacturing process. And by the good common sense of you, the model rocketeer, in using the product safely. It's worked. Years ago, many people didn't think it would. But the NAR, the manufacturers, and millions of model rocketeers proved it would and does.

As stated earlier, thrust and time (duration) are equally important in determining model rocket motor performance. If you were to use a model rocket motor with a thrust of 100 pounds (445 newtons)—if you could buy one, which you can't because average thrust of model rocket motors is limited to 80 newtons (17.98 pounds-force)—you'd never see even a 16-ounce model rocket leave the launcher! Likewise, a thrust of only 0.1 pound (0.445 newton) lasting for 10 seconds probably won't even lift a model rocket into the air. The performance factors of a model rocket motor that *really* determine the flight characteristics of a model rocket are thrust, duration, and their combination, called *total impulse.*

It was stated above that, roughly speaking, total impulse is determined by multiplying thrust by duration. However, thrust isn't constant. Therefore, total impulse is usually determined by a more precise method: measuring the area under the thrust-time curve, which amounts to the same thing. For math buffs, we integrate the area under the thrust-time curve.

Using a thrust-time curve generated during a static test, we can determine this area by laying the recorded chart down over a sheet of quad-ruled paper and counting the number of squares inside the thrust-time curve. That's the difficult way to do it. A more accurate result will come from using an instrument called a *plane polar planimeter* to measure the area under the curve. However, these manual methods have been superseded by electronic means using computers to digitize the data and carry out the mathematical integration.

The result of measuring the area under the thrust-time curve to determine total impulse comes out dimensionally in terms of thrust times duration—newtons × seconds or newton-seconds. In the old English engineering system, which is not easy to use in model rocket performance calculations, this would be in terms of pound-seconds.

Total impulse roughly determines how fast and how high a model rocket will fly. Flight calculations discussed in a later chapter will show why this is so. For now, it's enough to know that a motor with higher total impulse will propel a model rocket to a higher altitude. That is why the NAR rates model rocket motors in classes based on their total impulse. The NAR total impulse classifications are shown in Table 6. All U.S. manufacturers use these classifications.

TABLE 6
Model Rocket Motor Total Impulse Ranges

Type	Newton-Seconds	Pound-Seconds
$\frac{1}{4}$A	0.000 to 0.625	0.000 to 0.14
$\frac{1}{2}$A	0.626 to 1.25	0.15 to 0.28
A	1.26 to 2.50	0.29 to 0.56
B	2.51 to 5.00	0.57 to 1.12
C	5.01 to 10.00	1.13 to 2.24
D	10.01 to 20.00	2.25 to 4.49
E	20.01 to 40.00	4.50 to 8.99
F	40.01 to 80.00	9.00 to 17.98
G	80.01 to 160.00	17.99 to 35.95

Thus, the *first letter* in the NAR motor code tells you the total impulse range of the motor. Most motor types perform about 10 percent below the top of the range to allow for statistical variation. However, some of the larger model rocket motors may fall in the middle of the range. You don't need to know the precise total impulse of a model rocket motor to get some idea of its general performance, because each total impulse class is roughly double that of the previous class. A Type B motor will have twice the total impulse of a Type A motor, and a Type C motor will have twice the total impulse of a Type B and four times that of a Type A.

The first *number* in the NAR model rocket motor code (Type B6-4 in our example) tells you the *average thrust* of the motor in newtons. This is a derived, or calculated, number. It's determined by dividing the total impulse by the duration. It indicates what the thrust *would be* if it were constant from ignition to burnout. Average thrust is a very useful piece of information for altitude prediction. It also gives you an indication of how much your model rocket will accelerate and roughly how fast it will be going when it leaves the launch rod. We can clarify this by looking at Figures 5-9, 5-10, and 5-11.

Figure 5-9 shows the *idealized* thrust-time curve of a *constant-thrust* hypothetical Type B10 motor with an average thrust of 10 newtons for a duration of 0.5 second. This is a high-acceleration motor useful for flying heavy models or for flying regular models in high winds to prevent weathercocking; more about this later.

The hypothetical Type B5 motor whose thrust-time curve is shown in Figure 5-10 has an average thrust of 5 newtons and a duration of 1.0 second. The Type B5 motor has the same total impulse as the Type B10 motor. But the Type B5 would be a better choice for a small, lightweight, high-altitude model.

The Type B5 and the Type B10 have the same total impulse—5.0 newton-seconds (abbreviated n-sec). But their thrust characteristics are quite different, and they would make the same model perform quite differently.

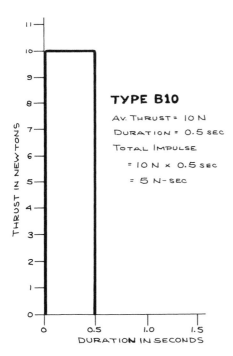

FIGURE 5-9   Thrust-time curve of a hypothetical Type B10 model rocket motor.

Figure 5-11 shows the thrust-time curves of two motors with identical average thrusts but different total impulses. The Type A5 has 5 newtons of thrust and a duration of 0.5 second. The Type B5 also has 5 newtons of thrust but a duration of 1.0 second. Which motor will take the same model to a higher altitude? Obviously, the Type B5 will, because thrust will be applied to the model for a longer period of time.

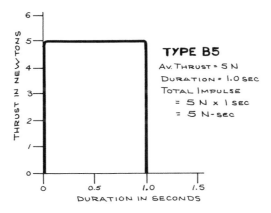

FIGURE 5-10   Thrust-time curve of a hypothetical Type B5 model rocket motor.

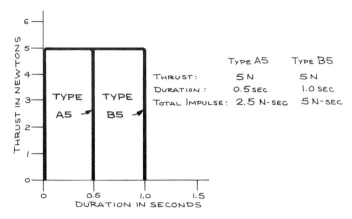

FIGURE 5-11   Thrust-time curves of hypothetical Type A5 and Type B5 model rocket motors showing equal thrusts but different total impulses.

You can get a rough idea of a model rocket motor's duration by dividing the maximum total impulse of its motor class by its average thrust number. The *number following the dash* in the motor code tells you the number of seconds after burnout before the time delay charge activates the ejection charge and deploys the recovery device.

Thus, in our example, the Type B6-4 motor has a total impulse range of 2.51 to 5.00 n-sec (probably being 4.90 n-sec at the top of the range), an average thrust of 6 newtons (about 1.35 pounds of thrust), and a time delay that will activate the recovery device 4 seconds after burnout.

This is nearly everything you need to know in order to select the proper model rocket motor for your model rocket. You can't get by with any less information than what's in the NAR motor code. The system has been in use since 1960. There have been many attempts to simplify it, but nobody has yet been able to devise a coding system that's so easy to remember but still tells all the basic information you need to know about a motor. If anybody says the system is too complicated to learn, it's my guess that that person doesn't have enough gray matter to get safely involved in model rocketry in the first place.

You must always match the motor to the model *before* flight. You must select a motor with enough average thrust to accelerate your model safely into the air; the heavier the model, the higher the average thrust should be. Most importantly, you must select the proper time delay. In the beginning, *follow the recommendations of the kit manufacturer regarding time delay.* Later, you'll have enough experience to decide for yourself which time delay to use.

If the time delay is too short, the model rocket will deploy its recovery device while it's still climbing. This may happen at high airspeed, and the aerodynamic forces may tear the recovery device to shreds or rip it entirely off the model.

If the time delay is too long, the model will climb to apogee, pause, arc over, and begin to fall back toward the ground. It will gain speed as it falls. Again, the recovery device may be deployed at high airspeed and tear itself to pieces. The model itself may be destroyed in the process. These "cliff-hanger" flights are no fun and could be hazardous.

Rule of thumb: When in doubt, use a *shorter* time delay. If the recovery device deploys while the model is still going up, use a longer time delay motor on the next flight.

Some model rocket motors are dash-zero types—that is, the motor code ends in a zero (Type B6-0, for example). These motors have no time delay or ejection charge. They are intended for use in the lower stages of multistage model rockets (see chapter 11).

Another rule of thumb: Don't overpower your model. First flights should be made with the lowest total impulse motor type recommended by the kit manufacturer. If you overpower your model rocket, you're likely to lose it on the first flight. The kit manufacturer and hobby store owner will love you because you'll have to buy another kit. The motor manufacturer will love you, too, because motors of higher power contain more propellant and therefore cost more money.

Most model rocket kit manufacturers show the altitude that their kits will attain with various motor types. However, without mathematics and computer programs, you can estimate how high your model will go. As we'll see in a later chapter, it's possible to predict with great accuracy the altitude a model will achieve, but that's a bit more advanced. For starters, you can expect the following performance from a basic beginner's model weighing about 40 grams (1.4 ounces) without a motor, using a 25-millimeter-diameter (0.984 inch) body tube, and equipped with these motors:

Type A6-4	250 feet (76.2 meters)
Type B6-4	450 feet (137.2 meters)
Type C6-5	1,000 feet (304.8 meters)
Type C6-7	1,100 feet (335.3 meters)

Thus far, we have discussed only the smaller types of model rocket motors that are factory-loaded expendables. Most model rocketeers fly with motors from Type A to Type C. All of these, with one or two exceptions, look like the cutaway drawing of Figure 5-3 and produce a thrust-time curve similar in most respects to that in Figure 5-7. This type of motor is categorized in rocket propulsion terminology as a *restricted end-burning grain* design. Some professional rocket engineers call it a cigarette-burning grain because burning starts on one end and proceeds to the other as a cigarette does.

A solid rocket propellant charge, or grain, burns only on the surface that is exposed for burning. Therefore, it produces an amount of hot gas in proportion to the amount of propellant area that's burning at any given instant. To get more gas and therefore more thrust, a rocket engineer will design a propellant grain so that more burning area is exposed or tailor the thrust level by

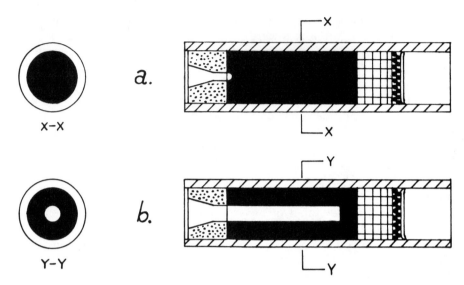

FIGURE 5-12   Cross-section diagrams of hypothetical model rocket motors showing (a) typical end-burning propellant grain, and (b) typical core-burning propellant grain.

controlling the amount of area exposed at any given time. This is done by shaping the grain into different configurations.

A wide number of grain configurations are used in professional rocketry. For example, a space shuttle solid rocket booster (SRB) has a star-shaped hole up its center; it also has different grain configurations at different locations in the motor to cause more or less propellant area to be exposed for burning at any given time during operation. This effectively throttles the SRB thrust, keeping it at a level that will provide a comfortable ride for the Orbiter and its contents.

Three basic grain configurations are used in model rocketry. The first, which we've discussed, is the restricted end-burner. The second one is used for small, high-thrust, 18-millimeter motors and for some of the larger Type E, Type F, and Type G motors. It is an *unrestricted core-burning grain* configuration such as is shown in Figure 5-12.

Since the end-burner has a constant burning area exposed at all times (except at ignition when it is a modified core-burner for high initial thrust), the end-burner produces almost constant thrust.

With its hole up the middle, the unrestricted core-burner grain has more propellant area exposed for combustion. With this greater burning area available, this grain design produces more thrust. It burns not from one end to the other but *outward* from the center. The thrust *increases* during burning as more propellant area is exposed. See Figure 5-13.

The thrust-time curve of an end-burner isn't optimal for model rocket use, where we need a lot of thrust at the start and less thrust as the model grows

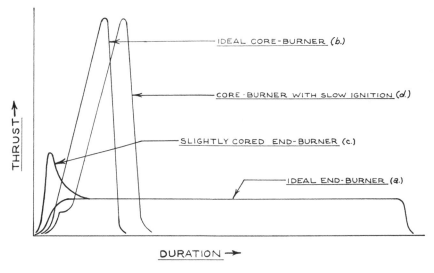

FIGURE 5-13   Thrust-time curves of various model rocket motor types with differing propellant grain configurations.

lighter because the propellant burns away. Therefore, a third type of grain called the *C-grain* has been developed. This is basically an end-burning grain that has a slot cut lengthwise along the propellant charge. This provides more burning area than an end-burner and also burns in such a way that the thrust-time curve is very much like that of an end-burner. C-grain designs are used mostly in model rocket motors that use the advanced composite rocket propellants.

NAR standards require that a model rocket motor manufacturer publish the thrust-time curves of the motor types the company manufactures and sells. The instructions that come with all model rocket engines also include directions on how to safely ignite the motor. Again, *read and follow all instructions carefully.*

As you can now see, a model rocket motor appears to be a simple device and yet is really somewhat complicated. It is the result of decades of development of solid-propellant rocket motors by professional rocket engineers and model rocket motor manufacturers. Each model rocket motor type took months of work to bring it up to the level of performance and reliability that would permit it to be NAR-certified and sold to the public. Many thousands of dollars were required to make the special machines and equipment necessary to manufacture such reliable model rocket motors. Often, this equipment is hazardous to operate. A great deal of credit must be given to the model rocket manufacturers who undertake grave risks to make that simple model rocket motor. It's packaged Space Age power. Treat it properly and it will perform with great reliability.

6

IGNITION AND
IGNITION SYSTEMS

To fly a model rocket properly and safely, you must have an electrical ignition system and launch pad.

The need for an electrical ignition system and a proper launch pad comes from years of experience. I was trained in the safety and handling of explosives, rocket motors, and pyrotechnics at White Sands Proving Grounds (now White Sands Missile Range) in New Mexico many years ago. What I learned hasn't become trivial because of technological progress. I had good teachers—for example, people who had worked for Dr. Robert H. Goddard, the American rocket pioneer, as well as one Navy explosives expert whose nickname was "Boom-boom." Some of these people had learned the hard way and had lost some body parts in the process. If you do things right and heed the advice of those who know, you, too, can proceed through life with all your parts and faculties intact. That's what this book is all about.

An electrical ignition system and launch pad are what is known in professional rocketry as ground support equipment, or GSE. They are also called capital equipment items because unlike a model rocket motor they're used over and over again for many model rocket launchings. So, even though good GSE may cost $20 or more, you can relax in the knowledge that it's a one-time expense. You can use the ignition system and launch pad for flying thousands of model rockets if you get good equipment to start with and maintain it properly.

I've used the same launch pad for as long as twenty years. I replaced it not because it was broken or worn out but because I wanted to upgrade to better equipment as it became available. I built an electrical ignition system back in 1962 before such a thing was available in the hobby shops. I still use it, although I've replaced the internal 12-volt battery several times. I also have several of the simpler manufactured ignition systems that I use when I don't need the heavy-duty older system.

Today, all major model rocket manufacturers sell one or more electrical ignition systems and launch pads. Most of the simple units available from

FIGURE 6-1   A model rocket is always launched by controlled electrical means using a launch pad and an electrical ignition system.

Estes and Quest are preassembled and ready to use. I highly recommend that you purchase these products, although you may be perfectly capable of making an electrical ignition system yourself. The manufactured units contain all the necessary and desirable safety and operational features. They may actually cost less than if you purchased all the parts separately and built your own.

And although you may purchase your model rocketry GSE, I'm going to describe how to make a simple electrical ignition system and an inexpensive launch pad. This will help you understand your commercial GSE so that you can troubleshoot and fix it should something go wrong.

This chapter will cover ignition technology and systems—how to get the model rocket motor ignited in the first place. The next chapter will be devoted to launchers and launching techniques.

Let us begin with a very basic assumption: when they launch a space shuttle at Cape Canaveral, they don't light a fuse and run away. They do it electrically.

They either push a button or depend on a computer to do it—but they always have the capability to override the computer. Moral: always have an EXIT command, STOP switch, or way out.

Model rockets are always ignited by remote electrical means. Electrical ignition is simple, reliable, and safe. In the past ten years, ignition technology has given us much better electric igniters than we ever had before.

It's not only unlawful in every state to attempt to ignite a model rocket motor with a fuse, it's also very dangerous. You must not do it! So that you'll completely understand why, here are some reasons:

1. Fuse ignition is not reliable. Some kinds of fuses will not fit into a model rocket motor nozzle and therefore will not ignite the motor. Some model rocket motors cannot be safely ignited except with the electric igniter that is designed for the specific motor type and is installed according to instructions. It's also possible for a fuse to have a hangfire, to borrow an old gunnery term. In this case, remnants of the fuse may smolder for more than 30 minutes up in the nozzle where you cannot see; you don't know when, how, or if ignition is going to take place. And you don't dare go out to the model to find out!

2. Fuses cannot be accurately timed. The sort of fuse that can be bought in some places isn't reliable. The fuse could flare up in a fraction of a second, and the model could take off in your face. Never, never, *never* trust the burning rate of a fuse—not when your safety and that of other people depend on it.

3. Fuse ignition gives you absolutely no control over the moment of launch. You cannot stop the liftoff in the last split second if you have to—not with a sputtering fuse on the launch pad under the model. I've actually seen the following events take place in the last 5 seconds of model rocket countdowns: (1) a low-flying airplane (which shouldn't have been there anyway but was) appeared over the crest of a hill and flew directly over the launch site at an altitude of about 100 feet; (2) somebody got excited and ran out to the launch pad, completely ignoring the shouts and screams of the range safety officer; (3) a strong gust of wind blew the launch pad over, so the model was pointed directly at me; (4) an unsupervised child spectator ran into the launch area chasing a ball; and (5) the clown hooking up his model on the next launch pad wasn't paying any attention to what was going on and backed away from his launcher directly into the launch pad and model that was in the final countdown stages. Because we were flying by the rules and using electrical ignition, the countdowns were stopped immediately and safely. They could not have been stopped if a fuse had been sputtering on the launch pad.

4. With fuse ignition, glowing remnants of the fuse are ejected from the model rocket motor nozzle upon ignition. These can fall into dry material around the launch pad and start a fire. I wasn't there, but I

know of at least one case where a sputtering fuse fell out of the motor nozzle, landed in dry grass around the launch pad, and started a significant fire that required the fire department to put it out. In general, fire marshals and fire chiefs are now friendly to model rocketry because many of them were once model rocketeers themselves. And they're friendly to model rocketeers if the rules are followed. But they don't like fuses, and who can blame them?

5. At Cape Canaveral, they do not believe it's safe to ignite the space shuttle engines with fuses. When professional rocket engineers go back to lighting fuses and running for the safety of the blockhouse, we model rocketeers *might* reconsider our requirement for electrical ignition—maybe.

6. Fuses are for fireworks, not model rockets.

Simple, reliable, and safe electrical ignition gives you the opportunity to stand at a safe distance from your model rocket and send it on its way with a professional countdown that gives you complete control up to the instant your finger comes down on the launch button. If you have a misfire—and we've all had them, even with the most careful preparation—you can disarm the electrical circuit with better than 99 percent confidence that the model rocket isn't going to ignite and lift off.

Every model rocket motor sold today comes with an electric igniter and explicit, complete instructions on how to use that igniter properly. Although I'll discuss electrical ignition in general terms here using a generic electrical igniter as an example, you must always *read and follow* what the motor manufacturer says about safe electrical ignition. Not all model rocket motors can be safely and properly ignited using the general methods described herein. Often, special igniters and methods must be used to ensure fast, reliable motor ignition. This is especially true of the larger model rocket motors using composite propellants.

All electrical igniters are based on the principle of the electric heating element. The way a hot wire works is easily understood. All matter has some resistance to conducting electrical current. Poor conductors are called *insulators*. Materials that conduct electricity easily are called *conductors*. Both insulators and conductors have some *resistance* to electrical current. When electrical current encounters this resistance, some of its energy is converted from electricity to heat. Push enough electrical current through any material, and its resistance will cause it to get warm or hot.

An igniter operates sort of like trying to push 300 people down a big hallway and through a single revolving door in a hurry. The resistance to the movement of the crowd through the restricted opening causes tempers to heat up. The energy of crowd movement is converted into the energy of confrontation.

The basic element of an electric igniter is a short piece of high-resistance material that will get hot because of its electrical resistance. The heat element of an igniter is usually made from nickel-chromium, *nichrome*. It's the same wire that's used in an electric toaster. And it works the same way. When an electrical current is passed through it, its electrical resistance and inability to pass the current causes it to get hot. Nichrome wire is made so that it won't melt or lose strength at red-hot temperatures, as would iron or copper. This hot wire in contact with the propellant either raises the propellant to ignition temperature or activates other materials in the igniter itself.

An igniter that uses a hot wire to cause other pyrotechnic materials to flare up and in turn ignite the propellant is called a squib igniter, or simply a *squib*. The Estes Igniter has such a material on its tip to help ignite the model rocket motor.

The Tiger Tail and Copperhead igniters from Quest and AeroTech, respectively, don't use hot wires. The black tip of these patented igniters is a squib composition containing fine carbon fibers. Carbon will conduct electricity and can be engineered to have the sort of electrical resistance necessary to serve as the hot wire. Early electric light bulbs had carbon filaments.

The ignition temperature of most model rocket motor propellants is in excess of 550 degrees F (288 degrees C). Some composite model rocket motor propellants require both high temperature and high pressure to ignite because they will not burn at atmospheric pressure. Thus, a squib igniter must be used that produces both a flare of flame and a brief pulse of pressurized gas from squib material combustion.

The basic hot-wire igniter is shown in Figure 6-2. It will work with most model rocket motors from Type ½A to Type D that use blackpowder propellant. A 2-inch length of #32AWG (American Wire Gauge) nichrome wire is doubled over and inserted up the nozzle until it is in contact with the propellant. If the wire isn't in contact with the propellant, it won't ignite the propellant. It is held in place with a tiny wad of flameproof recovery wadding stuffed into the nozzle, with paper tape over the end of the motor, or with a plastic igniter tack that fits into the nozzle.

Don't worry about the wadding or igniter tack. Once the propellant starts burning, the internal pressure of the motor quickly builds up to more than 100 pounds per square inch (6.8 atmospheres). Both the igniter wire and the wadding or tack will come out of the nozzle in a hurry.

The nichrome wire must be of very small diameter in order to possess high electrical resistance. #32AWG nichrome wire is only 0.00795 inch in diameter.

Heating the wire requires both electrical current and electrical voltage or pressure to push the current through it. The amount of electrical current that will flow through any conductor is determined by Ohm's Law, which is written

$$I = E/R$$

where E = voltage in volts, R = the electrical resistance in ohms, and I = the current that will flow in amperes. The higher the voltage, the more current will flow. The higher the resistance, the less current will flow.

But it takes electrical *power* to cause an igniter to get hot. In an electrical circuit, the power is determined by the equation

$$P = I^2R$$

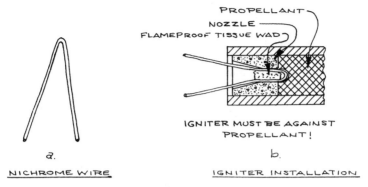

FIGURE 6-2   Drawing of a typical nichrome hot-wire igniter and its installation in a model rocket motor. This igniter type requires more electrical current and a bigger battery than the high-tech igniters now available, but its operation is the same.

where I = the current in amperes, R = the resistance in ohms, and P = the power in watts. This equation tells us that if you double the resistance, the power or energy that's dissipated in the resistance element (the hot wire) is doubled. But if you double the current in amperes, the power required goes up four times.

Together, these two equations tell us several things. If the voltage is too low, not enough current will pass through the igniter to generate enough power to raise the wire temperature to the point where the solid propellant will ignite. However, if you double the voltage, twice the amount of current will flow, and this will increase the power and the heat by a factor of 4.

The electrical resistance of 2 inches of #32AWG nichrome wire is about 1 ohm. If you use a 6-volt battery and a good electrical system that itself has a resistance of 1 ohm, 16 watts of power will get to the igniter. This is just about enough to get a #32AWG nichrome wire hot. However, if you use a 12-volt battery, it will have to deliver only ¼ the amount of current to achieve the same igniter temperature.

I once tested some igniters made from 2-inch lengths of #30AWG wire, 0.010 inch in diameter. I hooked some precision electrical measuring equipment into the circuit. I found that 5 volts is barely enough to get the wire hot; 6 volts did a better job and the wire glowed a dull red. At 12 volts, the wire glowed bright red. At 18 volts, the nichrome wire burned in two before it got red hot.

Never plug an igniter into a 120-volt AC house circuit. This is a direct short circuit. Under these circumstances, the nichrome wire will pass 90 amperes of current. This will probably cause the circuit breaker to open and raise havoc with your firing system. The nichrome wire will simply vaporize instantly without getting hot enough to ignite a model rocket motor.

It took many years to develop cheap and reliable model rocket igniters that would divorce model rocketry from the automobile battery that had to be used in the early days. The Estes Igniter was the first. It has a short piece of high-resistance nichrome wire spot-welded across the ends of two heavier lead wires with the tip dipped in squib material and the lead wires held by tape. The Quest Tiger Tail and AeroTech Copperhead igniters use a strip of plastic insulator material with conducting copper foil on both sides. The tip is dipped in a special composition that is a patented and proprietary mixture of carbon fibers to create a high-resistance bridge across the tip and squib material to produce a quick flare to ignite the motor. Each igniter has its fervent supporters and users; a good way to start an argument on the flying field or in a club meeting is to claim one is better than the other.

These commercial model rocket igniters all work very well with the 6 volts available from four fresh AA alkaline penlight batteries. They also work with 12 volts. You should be able to get several dozen launches from a set of AA alkaline penlight cells. Don't use ordinary penlight cells because they don't have the ability to deliver the required current without their voltage collapsing. More about battery behavior in a moment.

Available light-duty commercial electric launch controllers that use 4 AA alkaline cells—the Estes Electron Beam and the Quest Launch Controller— will ignite the Estes Igniter and Tiger Tail igniters. Don't think of either of them as cheap plastic toys. Both will last for years if a modicum of care is given to them.

Other electric launch controllers are available from many manufacturers. Usually, these are more complicated and cost more. All of them do the same thing: switch enough electrical current to a model rocket igniter to cause the motor to ignite, and keep electrical current from getting to the igniter when you don't want it (like when your fingers are underneath the model hooking up the igniter). Some units beep. Others have heavy-duty internal batteries. Some include capabilities for remote relay launching (more about this later).

Please do not buy, build, or use any electric ignition system that is (1) radio controlled (RC) or (2) has an automatic countdown without being sure it has some important safety features. The radio-controlled unit could be activated by someone flying his or her RC model airplane nearby or even by a trucker's CB rig on the highway a mile away. It's possible to have a safe RC launch system if it requires a combination of coded signals to activate it rather than simply a carrier signal.

The automatic countdown system is basically an electronic fuse that you start 5 to 10 seconds before liftoff and maybe can't stop. An automatic countdown system should have a fail-safe feature euphemistically called a "deadman switch," which is a switch that must be held shut in your hand to keep the countdown going; if you release it, the countdown stops instantly.

Remember: *Safety is no accident!* Try to keep in mind all the things that could go wrong. Sooner or later, one of them will.

I'm going to explain how to build an electric launch controller here so that you'll understand all of the safety features and operational requirements that exist, even in the simple Estes and Quest launch controllers that operate on the same principles with the same basic wiring. A properly designed and carefully

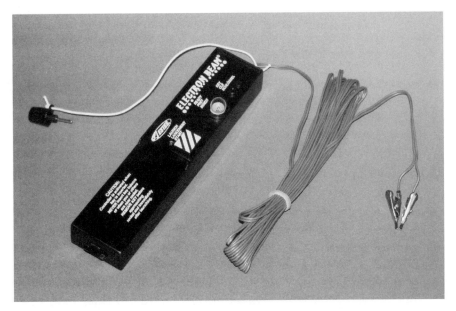

FIGURE 6-3   A typical commercial electrical launch controller with internal size AA alkaline penlight batteries for use with Estes and Quest igniters. Note the safety key and continuity check light.

constructed launch controller will last a long time. The basic components of an electric launch system are a launch controller, a battery, and connecting wires.

Basically, the launch controller is a switching device that makes sure no electrical current gets from the battery to the igniter before you want it to do so. Then, when you complete the prelaunch program, it delivers enough electrical current to the igniter to start the model rocket motor.

Its basic part is a spring-loaded push-button switch that will remain in the "off" or "open" position and will automatically return to this condition when released. In electronics and electrical engineering, it's known as a momentary contact normally open switch. An ordinary electric light switch or a *knife* switch will not make this automatic return to off, so these switches are extremely hazardous to use in your home-built launch controller. Sooner or later you'll forget to return the switch to the safe "off" position after a launch. The next time you put a model on the launch pad and hook up the igniter, the model will take off in your face. Having this happen to you once is more than enough, believe me! I've either made all the mistakes or seen them made. A simple ignition switch is an ordinary doorbell button as shown in Figure 6-4.

Safety rules also require a means to disconnect the battery from the system or a safety key that performs the same function. In addition to the ignition switch, the safety key in a launch controller physically opens the electrical

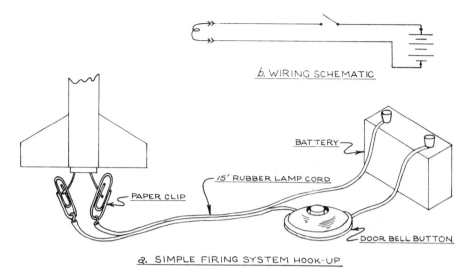

b. WIRING SCHEMATIC

BATTERY

15' RUBBER LAMP CORD

PAPER CLIP

DOOR BELL BUTTON

a. SIMPLE FIRING SYSTEM HOOK-UP

FIGURE 6-4   Construction and hookup of an inexpensive electrical ignition system. However, commercially available systems are inexpensive and are recommended.

circuit when it isn't installed. It provides a double precaution, something engineers call *redundancy.* When the safety key is taken out of the controller and is held in your hand or placed in your pocket, there is no way for any electricity to flow out to the igniter. Nobody is able to launch that model except you. Nobody can push the launch button just as you complete the igniter hookup. Although a safety key isn't shown in Figure 6-4 because safety means unhooking the battery with this simple circuit, you should have a safety key and *keep it with you.* Don't leave it in the launch controller. Don't keep it on a string tied to the launch controller. Put it on a long red or orange ribbon and put it around your neck. With the big, bright ribbon on it, you'll be able to look and see if you left it in the launch controller. The safety key is vital to your safety, and you shouldn't forget it for a single moment when you're on the flying field. With your safety key in your possession, you and *only* you can launch your model. You can work around the model on the launch pad knowing that nobody is going to play games with you and the launch controller while your fingers are under the model.

An ordinary automobile key ignition switch makes a good safety key unit. Once you have inserted the safety key into the launch controller, the controller is said to be *armed*—that is, it's ready to launch the model when you push the ignition switch.

A wiring diagram for a typical electric launch controller is shown in Figure 6-5. This is the basic circuit diagram for nearly all commercial electric launch controllers, too. You can see how the safety key opens the electrical circuit.

Notice the continuity light. This is wired across the ignition switch. It was invented in 1961 by Vernon D. Estes. It is a small light bulb that lights up

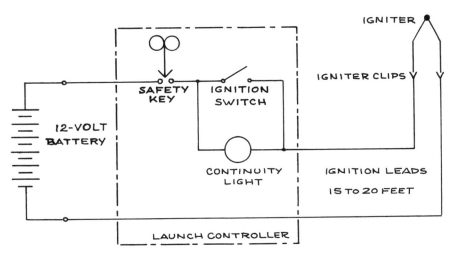

FIGURE 6-5   The electrical wiring diagram of a typical electrical launch controller showing the safety key and continuity check light. Nearly all commercial launch controllers are wired in this manner.

when you insert the safety key—if you've hooked up the igniter and everything is ready. The continuity light allows a small amount of current to pass through the circuit but not enough to get the igniter warm. It is a limiting electrical resistance. When the ignition switch is pushed and closed, the light bulb is bypassed by the low-resistance circuit of the ignition switch, allowing the full battery current to flow through to the igniter.

For a 6-volt ignition system, a Type 47 bayonet-base lamp bulb should be used; this will allow only 0.05 ampere to pass through the circuit. For 12-volt systems, use a Type 53 bulb.

Instead of a continuity light, some people use a piezoelectric buzzer that sounds off audibly. Radio Shack has a wide variety of these. Make sure that the unit draws less than 150 milliamperes (0.15 ampere) when it's working. The specifications on the package will tell you about this.

The current passed by continuity lights and buzzers is enough to activate some of the igniter types used in advanced rocketry. Such igniters use flashbulbs or sensitive electric match units. When you get to that point in model rocketry, *test your system ahead of time* to see if these advanced sensitive ignition elements fire when you arm the system with the safety key. I'm not covering those advanced ignition elements here because the technology is changing rapidly and may be different by the time you get to the point where you're using such things. For now, just be aware of these facts.

Except for electrical ignition systems used exclusively for the Estes Igniter or Quest Tiger Tail, the wire used between the launch controller and the launch pad should be #18AWG stranded copper. Don't use doorbell wire; it's too small and has too much resistance. Solid wire will break easily after being flexed a few times. Don't use loudspeaker wire, which is usually

#22AWG insulated stranded pair. What you want is commonly available #18AWG-2 insulated stranded rubber- or plastic-covered lamp cord, commonly called *zip cord*. Or you can cut up a long extension cord made of the same double wire. I prefer rubber-covered wire because rubber insulation flexes easily over a wide temperature range; plastic-covered zip cord gets stiff in the cold of wintertime flying.

The use of robust wiring should carry through the entire system. Heavy wiring and good connections reduce the electrical resistance of the system. Resistance causes the wires and connections to get hot. You don't want this. You want *the igniter* to get hot, not the firing system!

The NAR Model Rocket Safety Code requires that you be at least 15 feet from the launch pad if you're flying motors up to 30 newton-seconds (midrange Type E) and 30 feet when launching anything with more power than that. I suggest that you put a minimum of 35 feet of firing leads on your controller. This will allow you to fly everything up to and including a Type G motor without changing firing systems. Why the extra 5 feet? Over time, you'll be replacing the firing clips on the launcher end of the wires. The easiest way to do this is to cut the clips off, strip the wires, and install new clips. Over a period of time, the firing leads therefore grow shorter and shorter. So give yourself some extra to start with. The old advice of electricians should be heeded here: It's easy to cut off wire that's too long, but it's difficult to stretch wire that's too short.

You'll need something to hook the ignition leads to the igniter ends. Don't use big, heavy "alligator" clips. You *can* use ordinary, garden-variety metal *paper clips* as shown in Figure 6-4; they work fine, but you must solder the firing leads to them. (They don't work very well with Quest Tiger Tail igniters.) Commercial electrical firing systems come with little spring clips called *microclips*. Many companies make these. Ask for Mueller no. 34 Microgator clips or Radio Shack All-Purpose Micro Clips. Both of these have smooth jaws rather than the serrated or saw-toothed jaws of other clips. The smooth jaws won't pierce the copper foil of Quest Tiger Tail or AeroTech Copperhead igniters and cause them to short out. These microclips are cheap, so buy some spares because you'll have to replace them after a few hundred flights or so. They will become cruddy with exhaust residue, bent, and otherwise difficult to use. You can easily replace one in the field by cutting off the old one, stripping back the wire insulation about $1/8$ inch, and attaching a new clip by squeezing its tabs down on the bare wire with a pair of pliers. (You can solder it later when you get back to your shop.)

Before each launch session, test your electrical firing system to make sure it's working properly. Having a premature launch or a misfire is the hard way to test it. Use one of the spare igniters in your range box. Hook the clips to it *without putting the igniter in a model rocket motor*. When the safety key isn't installed in the controller, you should be able to push the launch switch without the igniter activating. Then put in the safety key. The continuity light should come on (or the continuity buzzer should sound). If it doesn't, something is wrong. Usually, it means dead batteries. Or the safety key isn't completing the circuit. If the light/buzzer comes on, then push the ignition switch. The igniter should fire when you do this. If the igniter doesn't activate, go into the troubleshooting mode.

The continuity light (buzzer) will *not* tell you that you don't have a short circuit between the clips or between the clips and any metal launch pad parts such as the jet deflector. You must check this visually when you're hooking up the igniter. The continuity light *will* tell you by failing to light up that you have dirty clips, a dead battery, or more serious problems inside the launch controller.

Here's a quick tip that makes using Quest Tiger Tail igniters faster and easier to use. Use a piece of tape—electrical or paper—to insulate *one jaw* of each microclip. Or you can slip a short piece of shrinkable plastic tubing over one of the jaws instead of using tape. When attaching these microclips to the foil of the Tiger Tail, place the insulated jaw of one clip on one side of the foil and the insulated jaw of the other clip on the other side of the foil. You don't have to use the Tiger Tail tape on the foil lead if you do this. This also works with the AeroTech Copperhead igniters, eliminating the need to use the special clip.

If you make your own electrical firing system, build it to last. Solder all connections if possible. Don't just wrap or twist the wires together. Remember: The system will have to pass enough electric current to operate a toaster or a 500-watt electric light bulb. Build it so that those little bitty electrons can get from the battery to the igniter with as little resistance as possible once you push the firing switch.

And *keep it simple!* Simplicity is safety and fewer things to break or go wrong. I have seen some incredibly complex systems over the years—ones that could talk, automatically count down, automatically check continuity, automatically stop the count if continuity wasn't there, and so forth. When they were working (most of them required a lot of tender loving care and maintenance), they did exactly what the simple systems do: ignite the model rocket motor. As you should know by now, I am a great advocate of the KISS principle—Keep It Simple, Stupid!

Batteries still give a beginning model rocketeer more trouble that anything else, even with a commercial electrical launch system that uses four AA alkaline cells. You can save yourself many of the headaches and heartaches of other tyro rocketeers if you'll take the time to read and heed what I have to say about batteries here. I've rarely had a launching battery fail me. In the few instances where it did, it was my fault. And I learned from that. So can you. Don't reinvent the wheel.

When you push the firing switch, you are, in effect, putting a dead short across your battery (or batteries; I'll use the singular to refer also to the plural hereafter). If you have a very good ignition system, its electrical resistance, including the igniter, may be about 2 ohms. To a 6-volt or 12-volt battery, this is indeed a short circuit. Ordinary flashlight dry cells—properly called LeClanche zinc-carbon cells—are not designed for this kind of treatment and are totally ruined immediately.

Alkaline dry cells will take it, however. Four AA alkaline dry cells are used in the Estes and Quest systems designed for their low-current igniters. But don't use them with the nichrome wire igniter described earlier as an example; these systems weren't designed to deliver the current to make such a primitive hot-wire igniter work. However, some model rocketeers have modified these commercial units to use a separate battery pack made up of four to eight size D alkaline cells, and these will last for about fifty flights. Other rocketeers have modified them for use with the 7.2-volt model race car and radio-control nickel-cadmium battery packs. These modified units will hack it, as the saying goes.

Next up the line in terms of size and power are the lantern batteries commonly found in the camping department of discount stores. They will fire low-current igniters for a long time and will last for more than a hundred flights firing simple hot-wire igniters.

All batteries convert chemical action to electricity. Like all chemical systems, they are dependent on temperature. When they get cold, they don't work as vigorously and won't deliver as much current. A dry-cell battery that works fine on a summer day may fail to fire a model rocket igniter in the dead of winter. Below 45 degrees F (7 degrees C), dry cells get sluggish and not very happy about supplying current for heavy-duty applications. So keep your launch battery warm in the winter until you get ready to fly. Keep it in the warm car until you're ready to put the model on the pad and hook it up. Another warming trick is to put the cold battery in your armpit inside your clothes, although it's something like putting an ice cube there. Some winter rocketeers are real hardy types, believe me!

Automobile batteries work fine for *all* sorts of model rocket ignition work. Some heavy-duty launch controllers come with big battery clips that will allow you to attach the firing system directly to a car battery. You don't have to remove the battery from the car; just connect these clips to the battery terminals. Or you can buy an adapter that plugs into the cigarette lighter.

Model rocketry was divorced from the car battery years ago, but many rocketeers still like to use these heavy-duty beauties because one will fire any igniter *right now* when you hit the launch button. Many rocketeers use small versions—motorcycle batteries that are miniature automobile batteries and are readily available at most auto supply stores at about half the price of the cheapest car battery. They weigh less than 5 pounds, are rechargeable, are easy to carry, and work great.

An automobile or motorcycle battery is technically known as a lead-acid battery. It uses a dilute sulfuric acid as its electrolyte and plates coated with lead paste. It has very low internal resistance, which means that it will deliver a lot of current without the voltage collapsing.

When you buy a new lead-acid battery, it is *dry-charged*, meaning that it doesn't have any battery acid in it. A container of acid is usually sold with a dry-charged battery. Battery acid is a mixture of sulfuric acid and distilled or deionized water and has a specific gravity of 1.280. It is put into the battery only once; thereafter, only distilled or deionized water (*never* ordinary tap water) is added as needed to maintain the proper electrolyte level.

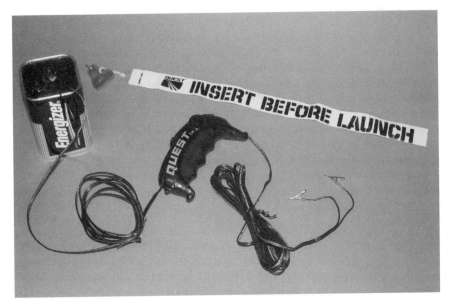

FIGURE 6-6   A Quest launch controller converted for use with a lantern battery.

When you put the acid into a new battery, *be careful.* The acid will make short work of your clothes if you spill any of it on them. If you get this acid electrolyte on your skin, wash it off *immediately* with lots of water. Wash down any electrolyte spills with lots of water. And don't tip the battery over or the acid may drip out.

All of this electrolyte mess and nonsense has been eliminated with the gel cell battery. This is a sealed lead-acid battery in which the electrolyte is in the form of a gel, which is a colloid, like gelatin dessert. Other features built into the gel cell eliminate the need to vent it during charging—if the charging current is kept within limits, as discussed below—and the need to replenish electrolytes. A 12-volt, 1.5-ampere-hour gel cell works beautifully for more than a hundred flights, although you should top the charge immediately after returning from the flight session. Such a gel cell is in a sealed plastic case about 7 inches long, $1\frac{3}{8}$ inches wide, and $2\frac{3}{8}$ inches high and weighs about a pound. I've got one built into my firing box along with the electrical circuitry. I recharge it after each flight session. I have to replace it every 5 years or so.

To really throw a heavy current to an igniter if you also want light weight, nickel-cadmium (ni-cad) batteries are the ultimate. They are also expensive. The cost has come down in recent years because of the increasing use of ni-cad batteries in portable power tools and hobby items. If you go the ni-cad route, get the high-discharge types because that's what you'll be using them for—high discharge rate, or throwing a lot of current through a model rocket igniter for a second or less.

All lead-acid, gel, and ni-cad batteries are rechargeable, but the charging conditions for each type and for each size are different. Charging instructions come with each battery. Follow these. Unless otherwise specified, the normal charging rate for all these batteries is $\frac{1}{10}$ of their ampere-hour rating. Thus, a 4.5 ampere-hour motorcycle battery should be charged at 0.45 ampere. Fast charging builds up heat in the battery and may warp the plates or dislodge active material from them, which will fall to the bottom of the battery and eventually short out the plates.

If you need a quick charge on the flying field and are using a 12-volt lead-acid, gel, or ni-cad battery, you can connect it to the car battery. I made up a charging lead with clips on one end and a plug that fits into the cigarette lighter on the other. A "buddy charge" for 5 to 10 minutes can usually give your weak battery enough zip to fire a couple of igniters.

The state of charge of a lead-acid battery can be determined by measuring the specific gravity of the electrolyte in each cell. You can buy an expensive battery hygrometer at an auto supply store to do this. When all cells show a specific gravity of 1.280, the lead-acid battery is charged. All lead-acid batteries produce gaseous hydrogen during the charging process, so charge them in a well-ventilated place such as an open garage. Keep the electrolyte level in each cell at the proper point by adding only distilled or deionized water.

A lead-acid battery left to itself will self-discharge after several months of inactivity. A lead-acid battery that has been completely discharged and will not accept a charge has become sulfated; its plates have become coated with a hard, insulating coating of lead sulfate. A sulfated lead-acid cell can sometimes be rescued by slow charging it for several weeks at a rate of $\frac{1}{100}$ of the ampere-hour rating. I've saved many "dead" batteries this way (including a couple of car batteries, which I then converted into launch batteries when I didn't mind hauling 25 pounds of car battery around).

Gel cells must be charged differently. Limit the charging current to one-fifth of the ampere-hour capacity until the charge voltage reaches 2.4 volts per cell (14.4 volts for a 12-volt battery). Then maintain a constant voltage until the charging current drops to $\frac{1}{10}$ of the ampere-hour capacity. Then disconnect because the charge is complete. A gel cell will hold its charge for several months.

Ni-cad batteries are sealed cells. You have no way to check the state of charge of these units. The best thing to do is to keep them on a constant trickle charge of $\frac{1}{100}$ of their ampere-hour rating at all times when you're not using them. Some people say you should let a ni-cad discharge completely before recharging it because trickle charging causes internal changes in the battery. However, I've used the trickle-charge method for thirty years with no trouble. Ni-cads will usually last for five years.

As you may have surmised from the above, I've built several battery chargers of my own that have voltmeters and ammeters mounted on them for monitoring the charge conditions. You don't have to do this anymore. You can buy battery chargers of all sorts now. Don't use a heavy-duty automobile battery charger on motorcycle batteries, gel cells, or ni-cads; that's like driving a nail with a sledgehammer. You can now buy little battery chargers specifically designed to charge the type and size of rechargeable battery you have. They cost about $20 and are a good investment.

Always watch polarity when you hook up any battery charger! Like everything else in model rocketry, if you do it right, you'll have years of success. And you'll rarely have an electrical ignition problem.

7

LAUNCHERS AND LAUNCHING TECHNIQUES

Good electrical ignition is only one part of getting a model rocket off the ground for a successful flight. You must also have a launch pad. A model rocket launch pad holds the model rocket vertically in a prelaunch position so that the igniter can be hooked up. Once the model rocket motor is ignited and begins to move upward, it also restrains the model's freedom of motion until it achieves enough airspeed to be stable.

If you set a model rocket directly on its fins on the ground with its nose pointed straight up and attempt to launch it in such a manner, it probably won't fly. (There are a *few* exceptions, but don't go looking for them.) It will probably lift off at an angle and chase you around the rocket range. Sitting on the ground, a model rocket has no air flowing over its fins to permit the fins to stabilize the model and keep it going in the direction it's pointed. About a quarter of a second is required for most model rocket motors to build up thrust and accelerate a model rocket to a speed of about 30 mph where air flowing over the fins can properly act to stabilize the model.

Without a launcher, a model rocket could topple over during the first split second of flight. If this happened, it could go all over the place. And you can't outrun it. For safety, a launcher must *always* be used!

A rod launcher is one of the oldest, simplest, and most commonly used launchers. Various styles are available at reasonable prices from model rocket manufacturers. They all have the following basic features:

1. A *launch rod* is nothing more than a piece of $1/8$-inch-diameter hard steel wire at least 36 inches long. It is held in a vertical or near-vertical position by the launch pad base. The $1/8$-inch by 36-inch rod is the NAR standard launch rod. In European countries, it is 3 millimeters in diameter by 1,000 millimeters (1 meter) long.

Most launch pad kits come with a two-piece rod. This allows the rod to be put into a shorter package. It's also a lot easier to carry around and store.

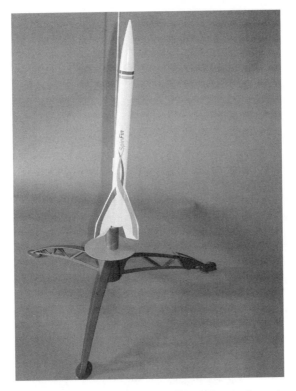

FIGURE 7-1   A typical commercial launch pad;
this one is from Quest Aerospace, Inc.

But with a two-piece rod, you're always in danger of losing or forgetting half of it. Half a launch rod will keep you from launching your models because you should *never* attempt to launch a model rocket with less than a 36-inch rod length.

If you want to make a two-piece rod into a one-piece rod, stick a ⅛-inch length of solder into the hole of one of the rods, and push the other half of the rod into the hole on top of the solder. Hold the assembly over a match or candle (being careful to keep your hands well away from the hot area) until the solder melts to hold the rods together. Or you can go to the hobby store and buy a 36-inch length of ⅛-inch-diameter "music wire," which is very strong steel wire used to make the landing gear of some model airplanes. The longer steel rod is clumsier to carry around, but you're always certain of having a full-length rod with you.

Launch rods get dirty and rusty. Put a wad of steel wool in your range box and use it to polish the rod before each flight session. Or you can pay a visit to your local steel supply warehouse (look in the telephone yellow pages) to buy a rod made of stainless steel that will never rust.

As a matter of fact, if you want launch rods of greater diameter and length for launching larger model rockets, you can buy these at most steel supply warehouses, too. I have a 5-foot length of $^3/_{16}$-inch-diameter rod and a 6-foot length of $^1/_4$-inch-diameter rod. I use an AeroTech Mantis launch pad that will accommodate these larger-diameter launch rods.

2. A simple but effective *launch pad base* is nothing more than a piece of $^3/_4$-inch plywood about 12 inches square. Drill a $^1/_8$-inch-diameter hole vertically into the middle of the board. Tap the launch rod down into the hole.

Most sophisticated launch pads, including some now available from manufacturers, are made with three or four legs. A good launch pad base will have its legs spread widely apart for stability. When the launch pad sits on the ground with a large model on it, it should be very difficult to tip it over. This is important when launching on uneven ground or on gusty days.

A launch pad should have a tilting mechanism so that the launch rod can be tipped slightly away from the vertical position. This allows you to put in a little bit of tilt to compensate for effects of wind, primarily weathercocking of the model when it takes off in a wind. We'll discuss this in detail later. The NAR Safety Code says that you must always launch with the rod pointed within 30 degrees of the vertical. All commercial launch pads are designed so that this limit cannot easily be exceeded. Tilting is accomplished by various means, depending on the design of the launch pad. If the launch pad doesn't have a tilt mechanism, you can always tilt the entire pad and launch rod by putting a small rock or piece of wood under one of the legs.

Always make sure the tilting mechanism is securely tightened so that the model cannot tip over the launch rod with its own weight. Also make sure the launch pad cannot fall over with a model in place and ready for launch. If necessary on windy days or with heavy or long models, put one or two rocks or bricks on the base or the legs to hold the launcher in place. Some modelers make U-shaped staples of $^1/_{16}$-inch wire or from a coat hanger. They hold launch pad legs down by inserting the inverted U over a leg and into the soil.

3. A *jet deflector* must always be used on a launch pad to prevent the motor exhaust from striking the launcher or the ground. A jet deflector must be made from steel. In most kits, it's a steel stamping. Although a flat metal plate can be and is used as a jet deflector, a better jet deflector is a bent or angled piece of sheet steel that turns the jet at right angles and directs it away from the model and the launch pad.

A jet deflector is necessary at all times but especially if your flying field is covered with grass. It can prevent a grass fire being started by pieces of hot wadding and igniter debris that come out of the motor immediately after ignition.

A jet deflector can be cut from a can. Always use a steel can, not an aluminum one. Deflectors made from aluminum will not stand up to the temperature of the jet exhaust. A hole will be burned through an aluminum deflector after only a few flights. A steel deflector will last for a long time.

4. An *umbilical mast* is another launcher feature that's very helpful. This is a rod or dowel standing a few inches to one side of the launch rod. It has a

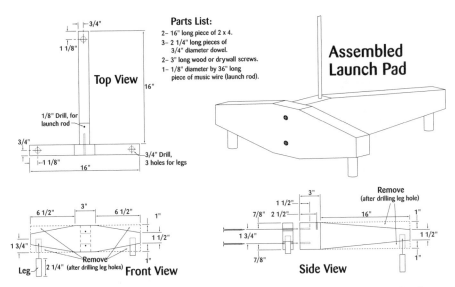

FIGURE 7-2 A simple launch pad that you can make from two 16-inch pieces of 2-by-4s. The 2-by-4s can be screwed together as a simple T, or you can make the extra cuts shown here and have a more realistic-looking pad.

clip, clothespin, or piece of tape for holding and supporting the weight of the ignition leads and clips. Some types of model rocket igniters have a tendency to be pulled out of the nozzle by the weight of the leads and clips when the igniter begins to get hot. Many misfires have occurred unnecessarily when the weight of the leads and clips pulled the igniter out of the motor before ignition could occur. An umbilical mast prevents these problems by supporting the major weight of the leads and clips, leaving only a few inches of firing leads for the igniter to support.

If I'm flying from someone else's launch pad that doesn't have an umbilical mast, I usually tape the ignition leads to the bottom part of the launch rod *below* the model's launch lug or to one of the launch pad legs. An umbilical mast is very important when launching front-motored boost gliders and rocket gliders, as we'll see later.

5. A *rod cap* should always be used to prevent accidental eye injury. Commercial launch pads come with rod caps. A rod cap can be nothing more than an old, expended motor casing painted a bright fluorescent color with a streamer or ribbon glued to it. The rod cap is placed over the top of the rod when the launcher isn't being used on the flying field or when it's being carried or stored. It alerts you to the location of the end of the rod so that you don't run into it. You can also make a rod cap by attaching a brightly colored cloth or plastic streamer to a spring clothespin and clamping it to the top of the launch rod when the launcher isn't in use.

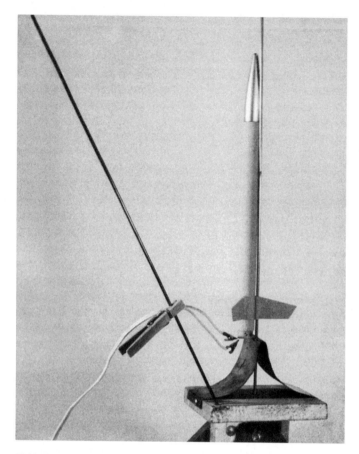

FIGURE 7-3   An umbilical mast and clothespin support the
weight of the electrical leads and prevent igniter pull-out.

You can further protect your eyes from the end of the launch rod and
from other things that might happen on the flying field by wearing sunglasses.
These are helpful whether the weather is bright or cloudy. Dark green or gray
lenses will help you see the model on a bright day, whereas yellow lenses are
extremely helpful on a dull, cloudy day.

Be sure to remove the cap from the top of the rod before launching your
model. One rocketeer put his model on the rod and attached a clothespin to
the top of the rod because he wasn't going to launch right away and he didn't
want somebody to walk into the launcher. Later he forgot to remove it. The
model stripped the clothespin off the end of the rod when it took off and
engaged in some rather spectacular aerobatics because the rod cap slowed the
model for a split second just as it was leaving the launch rod.

6. A *tarpaulin* or an *old blanket* should be used under your launch pad if
you fly from a dry, grassy field. This will not only keep the knees of your pants

dry and clean when you're hooking up your model but will also help prevent grass fires. The tarp or old blanket will catch any glowing pieces of wadding or igniter debris that might be ejected from the motor upon ignition. It's a precaution worth the trouble.

There are several helpful items you should have at the launch pad. Some of them can be tied to the pad with strings so that they don't get lost. One of these items is a *roll of paper tape* ³⁄₄ inch wide for use when you need to tape things up on the pad. For example, some models may require support on the rod to keep their tails from resting on the launch pad base. Tape can be used as shown in Figure 7-4 to hold a model up on the rod. But don't forget to give the model all the rod length possible; it will be more stable in flight if you do.

You should also have a small square of *no. 200 sandpaper* or an *emery board* such as that used to file fingernails. This is known as a JIC file (JIC stands for just in case). The file is used to remove the residue that forms on the igniter clips after several launchings to expose a clean metal surface and make a good electrical contact with the motor igniter.

I also keep a few *spring clothespins* in my range box or simply clip them to various parts of the launcher base where they'll be handy if I need them. They can be used in place of tape for supporting a model on the launch rod, for clamping ignition leads to the launch pad, and for handling the hundreds of other little clipping and clamping tasks that arise at the launch pad when you're hooking up models.

The simple, everyday things we model rocketeers use are often sources of amusement or consternation to nonmodelers. I'll not soon forget the time CBS TV News sent a team to cover a flying session of our model rocket club. The crew had just returned from Cape Canaveral and a major *Apollo* lunar mission. We put a 1:100-scale Saturn V on the launch pad, and I went underneath it to hook it up. TV cameras and microphones caught every breathless prelaunch action and word. Tension among the TV people was as high as it had been at the Cape. I turned to a fellow model rocketeer and said in a matter-of-fact tone, "Please hand me that clothespin." The TV crew, used to the intensity, drama, and sophisticated high technology of a full-scale Saturn launch, came completely apart with guffaws and laughter while the director yelled, "Cut! Cut!"

The NAR Safety Code requires that model rockets be launched from launchers tilted no more than 30 degrees away from the vertical. Why this limit? Answer: One word that you've read before—*safety.*

The reason can be traced to an Italian Renaissance artillery officer and mathematician, Niccolo Tartaglia (1500?–1557), who discovered and wrote down what any artillery officer, naval gunnery officer, or ballistician still calls Tartaglia's Laws of Gunnery. Tartaglia's Laws are as true for model rockets as for guns. Written in modern model rocket terms, they state:

1. A model rocket that will go 1,000 feet straight up will travel a horizontal distance of 2,000 feet if launched instead at an angle of 45 degrees.

FIGURE 7-4 A model can be held up off the launch pad base and jet deflector by a piece of paper tape wrapped around the launch rod as shown.

2. When launched at an angle of 45 degrees, a model rocket that would go 1,000 feet vertically will ascend to a peak altitude of only 500 feet during its arcing flight.
3. The path followed by a model rocket when launched at *any* angle other than the vertical will describe a *parabola*. (For very long-range shots such as an ICBM or a space vehicle, the path becomes an *ellipse* because for long ranges the curvature of the earth must be taken into account. See any physics book or a computer program for interplanetary trajectories of space vehicles.)

Tartaglia's Laws are correct for model rockets if you ignore air resistance. I tested them myself in the early days of model rocketry in the vast expanses of the desert Southwest under very careful safety controls (and before the

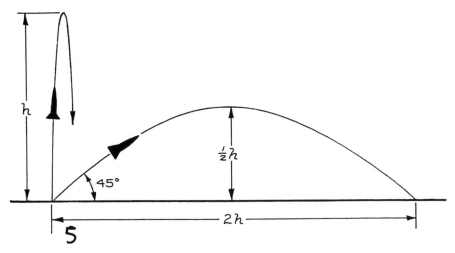

FIGURE 7-5 A graphic depiction of Tartaglia's Law. This is why the tilt of a launch rod is limited to 30 degrees from the vertical.

NAR was even founded and the NAR Model Rocket Safety Code developed). This is one experiment you shouldn't perform. There are many safer experiments you can do, so launch within 30 degrees of the vertical as the NAR Model Rocket Safety Code specifies.

To answer a question nearly everyone asks at first: no, a model rocket will not fly horizontally. It has no wings to provide aerodynamic lifting forces against gravity, and its fins won't do the trick because they're too small and too far aft. When launched horizontally, a model rocket falls off the end of the horizontal launch rod, flops to the ground, then skitters around on the ground during the thrust period of the motor. This performance usually removes the fins, bends or dings the body tube, and is not beneficial to a model rocket. It's not very healthy for human bystanders either because nobody has yet been able to outrun a model rocket.

The reason for permitting a launcher tilt of up to 30 degrees from the vertical is to allow the model rocketeer to compensate for wind effects on a model during flight. In a wind, a model rocket will exhibit launcher tip-off when the launch lug leaves the launch rod. The horizontally blowing wind makes the vertically ascending model rocket cock its nose *into* the wind like a weather vane or weathercock. That's why we call the phenomenon *weathercocking*. It happens with big NASA sounding rockets, too. The amount of weathercocking depends on the velocity of the wind, the weight of the model rocket, the stability characteristics of the model, the acceleration of the model, and other factors that would fill several pages with complicated partial differential equations.

But you don't have to know what a partial differential equation is to make weathercocking work for you. You can't fool Mother Nature, but you can usually make her work for you if you know how to do it. You can use the weathercocking phenomenon to make your model rocket go where you want it to go and do what you want it to do.

To gain *maximum altitude* during a model rocket flight in windy conditions, tilt the launch rod *downwind* or *with the wind*. The direction and amount of tilt is determined mostly by experience using the "wet finger" method of determining the wind direction and speed and a knowledge of how your model performs in a wind. When the model lifts off, the weathercocking effect makes it tilt *upwind* or *into the wind*. As a result, the model starts out at a tilt downwind but quickly weathercocks into going straight up over the launch pad. It will deploy its recovery device at maximum possible altitude, so you'd better have lots and lots of open area devoid of trees and houses in the downwind direction.

For *maximum duration* flights or when flying from small fields, tilt the launch pad *upwind* or *into the wind*. This will make the model weathercock even more strongly into the wind. It then flies upwind so that the recovery device opens well over the upwind side of the field. The model then drifts back over the field and launch site, and hopefully touches the ground just before reaching the rocket-eating trees on the downwind side of the launch field.

Most model rockets weighing less than 6 ounces (170 grams) can be launched safely from the standard ⅛-inch-diameter 36-inch launch rod. But when you start flying longer models such as the popular "superocs" that are "stretched" models ranging in length up to 6 feet, large payload-carrying models, models powered by Type E, Type F, and Type G motors, or large-scale models, you should use ³⁄₁₆-inch or ¼-inch rods 5 to 6 feet long. Heavier models require more time and distance to build up flying speed, and you should give them all the launch rod possible for the best guidance. Long superoc models may have their launch lugs as far up the body tube as 24 to 30 inches, which means that they've got less than a foot of standard-length launch rod above the launch lug for initial guidance.

There's absolutely nothing to be gained by using a short launch rod at any time, even though some ardent competition modelers claim that the friction of the launch lug on the launch rod decreases the maximum altitude of a model. Long launch rods will whip when the model is launched. Launching experiments made using high-speed cameras have shown that an aluminum launch rod will whip less than a steel one. (A ⅛-inch aluminum launch rod can be easily bent, which is why these small rods are made of steel.) Naturally, when you use a larger launch rod, you have to put a larger launch lug on your model to fit over the rod properly.

Another kind of launcher seen on some model rocket ranges, especially during NAR competitions, is the *tower launcher*. This is a structure, often very simple in construction, in which a model slides between three or four vertical guide rails. The use of a tower eliminates the need for a launch lug on the model. Experimentation has indicated that the aerodynamic drag of a launch lug can amount to 15 percent of the total drag on a model. Several manufac-

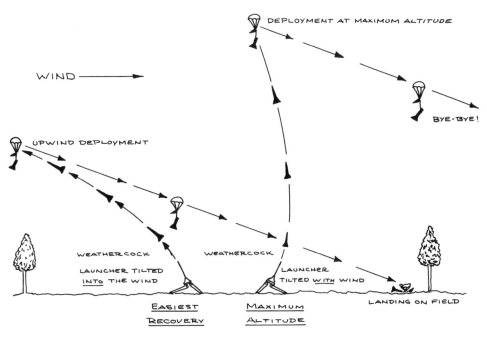

FIGURE 7-6   The launch rod can be tilted to make the model rocket fly where you want it to go and land where you can recover it.

turers such as Balsa Machining, whose tower is shown in Figure 7-7, make commercial launch towers.

There are many designs for launch towers, and they have both advantages and disadvantages. On the positive side, as was mentioned previously, a tower eliminates the need for launch lugs, which reduces the total aerodynamic drag on a model. A tower also provides a very stiff launcher that doesn't sway or bend in the wind or whip as the model is launched. On the negative side, a tower is a complex piece of ground support equipment (GSE) that easily gets out of adjustment, especially when you carry it around in the trunk of a car. The guide rails must be adjusted for each different model rocket body tube diameter. And some competition modelers actually believe (and some can produce data to support their point) that the friction of three or four tower guide rails is greater than that of a launch lug sliding over a launch rod—a factor they claim cuts down on the maximum performance to an even greater extent than the additional drag of the launch lug. (After all, model rocketry is a technology in miniature. In all fields of science and engineering, controversy exists. According to the famous aerodynamicist and rocket pioneer Dr. Theodore von Karman, "How can we progress without controversy?")

Tower launchers are built from wood, plastic, or metal. The one shown in Figure 7-7 is typical of the launch tower types now in use. Some are designed

FIGURE 7-7   A tower launcher such as this one from Balsa Machining can be adjusted to accommodate models of different diameters. A tower eliminates the need for a launch lug and is stiffer than a standard rod for launching heavy models.

to fit on inexpensive camera tripods so that the modeler can adjust the tilt as with a rod launcher.

A third kind of launcher is the *rail launcher*. It has also seen increasing use in model rocketry, especially with large and heavy models. In some cases, it has replaced the use of large launch rods entirely.

Another kind of launcher is the *split-lug* type shown in Figure 7-8. The launch lug itself is split to clear the support for the attachments of the rod to a rail launcher. A model designed for a split-lug launcher can also be flown from a standard launch rod.

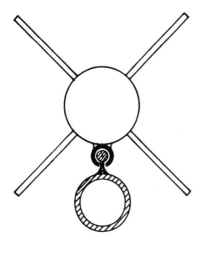

SPLIT-LUG

FIGURE 7-8 Cross-section of a split-lug launcher rail and model. A split lug can be made from a piece of thin-walled aluminum hobby tubing.

Since the early 1970s, competition model rocketeers have used a launcher design known as a *closed breech* or *piston launcher*. It's still a tricky launcher to use and I don't recommend it for beginners. Instead of allowing the exhaust from the model rocket motor to expand freely into the surrounding atmosphere, a closed breech or piston launcher uses the exhaust gas to pressurize an enclosed volume and move a piston to which the model is attached.

Early piston launchers consisted of an enclosed tube into which the model was slipped atop a wooden piston that would come off when the model left the tube. Howard R. Kuhn invented the most popular form of the piston launcher that is widely used in competition today. In this piston launcher, the pressure tube is a standard body tube that slips over the lower $1/4$ inch of the exposed motor in the tail of the model. The fit has to be *just right*. The piston and rod are attached to the launcher. When the motor is ignited (it's sometimes a tricky proposition to hook up an igniter and maintain hookup inside that enclosed tube), the exhaust gas pressurizes the tube and forces the tube and model upward against the piston. At the limit of piston-tube travel—about 12 inches—the tube comes up against a stop ring, the model pops off the top of the tube, and the bird is launched. Early piston launchers had no guidance means for the model after it popped off the upper end of the tube. Improvements to the design include 3-inch-long wooden guideposts on the launcher tube that bear against the side of the model and provide some lateral restraint. The action of a piston launcher is spectacular because the model goes "pop" and is in the air almost immediately. The pressure of the trapped exhaust gases in the piston launcher results in a very high initial acceleration.

Again, heed my advice and remember the KISS principle. It is as applicable to launchers as to ignition systems. Simple is better. Unless you are in

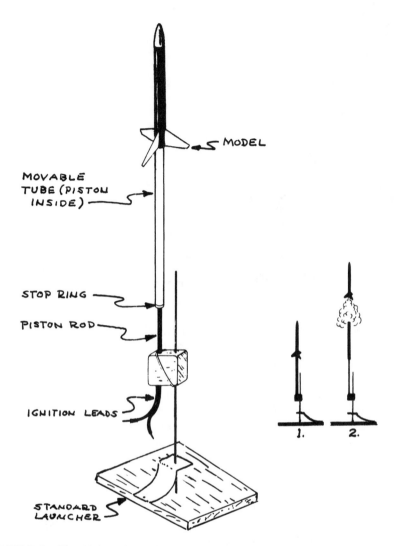

FIGURE 7-9 The Kuhn piston launcher uses the exhaust gas from the model rocket motor to pressurize a tube and piston for rapid acceleration of a model at liftoff.

national or international competition where a fraction of a percent of performance improvement might win, stick to the simple equipment. You should be interested in building and flying model rockets. However, I know that some model rocketeers like GSE and will spend a lot of money on it. Your GSE will be with you for a long time in your model rocket activities.

8

HOW HIGH WILL IT GO?

Once a model rocket leaves the launcher, it is a free body in space, even though it's still surrounded by the earth's atmosphere. It has been projected beyond the earth's surface, and its actions as a free body in space cannot be duplicated easily while it's on the ground. But we can account for the effects of the earth's atmosphere. If we subtract these effects, we can study the motion of the airborne model just as if it were in outer space. We can discover where it will go, how far it will go, and how fast it will go.

It is possible to fly a model rocket on paper. All you need are the elementary tools of simple arithmetic—addition, subtraction, multiplication, and division—and pencil and paper. Better have an eraser, too, because even the experts make mistakes. With these, you can find out in advance of flying your model exactly how it will perform. You don't have to be a genius to do this. And it's very exciting to work out all the numbers, then have the model perform the way the numbers said it would.

When the first edition of this book was published in 1965, hand calculators and home computers were nonexistent. Electronic digital computers were so big and so expensive that nobody but large corporations and universities could afford them. Personal computers have revolutionized all of the formerly tedious calculations involved with studying model rocket performance, designing model rockets, and determining model rocket altitude. Therefore, although we'll proceed through the basics of doing it by pencil-and-paper methods, we'll also discuss using some simple computer programs. Most of the tutorial in this handbook is based on the use of a program called RASP-93. There are also several other good programs available. Rocksim model rocket design and simulation software from Apogee Components is one of the most widely available programs out there. You can download a free demo version at www.apogeerockets.com.

The reason for discussing the pencil-and-paper methods is so that you'll get an understanding of what goes on during a model rocket flight. Based on that, you'll be able to work any bugs out of your own computer program.

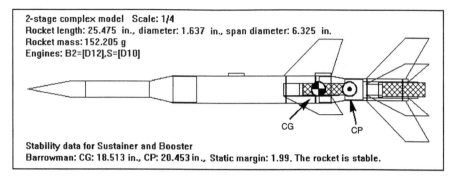

2-stage complex model Scale: 1/4
Rocket length: 25.475 in., diameter: 1.637 in., span diameter: 6.325 in.
Rocket mass: 152.205 g
Engines: B2=[D12],S=[D10]

CG   CP

Stability data for Sustainer and Booster
Barrowman: CG: 18.513 in., CP: 20.453 in., Static margin: 1.99. The rocket is stable.

FIGURE 8-1   Finding the center of pressure (CP) location for a complex design can be done quickly with a computer software program like Rocksim from Apogee Components.

(One of the disadvantages of the digital computer revolution has been the growing inability of people to recognize whether the computer answers are in the ballpark. Does the answer *look* reasonable? Is the decimal point in the proper place? These are questions that immediately come to mind among those who were forced to use pencil-and-paper calculations or to work with a slide rule. By starting with the basics, you'll be able to determine ballpark numbers in advance and make a pretty good SWAG—scientific wildly assumed guess—at whether your answers are basically correct.)

Although a model rocket has three flight phases—powered flight, coasting flight, and recovery—we're going to discuss only the first two phases here. Recovery will be treated separately in another chapter.

Three basic forces act on a model rocket in flight. You can think of a force as the application of energy to the model in such a way that the flight path is changed. As shown in Figure 8-2, and assuming *vertical* flight, these three forces are

1. *Thrust* from the model rocket motor that acts on the back of the model and makes it accelerate, or change speed.
2. *Gravity,* which is a force trying to slow the model in its vertical flight and acting in an opposite direction to thrust.
3. *Aerodynamic drag* caused by the model moving through the air. This drag force also acts to slow down the model.

During powered flight, all three forces act on the model. But during coasting flight, the thrust is zero; therefore, only gravity and aerodynamic drag act on the model.

The effects of aerodynamic drag are complex, as we'll see later. The effects of thrust and gravity are simpler to handle. So, to begin with, we're going to assume that there is no aerodynamic drag on the model—that is, the model is assumed to be a *perfect* aerodynamic shape, a fiction that, if real, would elim-

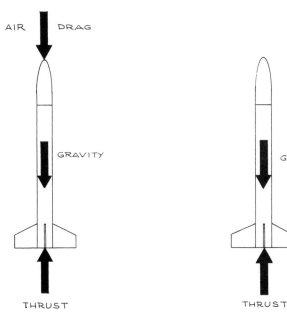

FIGURE 8-2   The forces acting on a model in vertical flight.

FIGURE 8-3   The forces acting on a model in the absence of air, which creates drag. This is a purely hypothetical situation.

inate a lot of scientific and technical headaches among professional rocketeers. So we're going to ignore the effects of the earth's atmosphere and *pretend* that our drag-free model is acted upon only by thrust and gravity, as shown in Figure 8-3.

By calculating the flight of the model as if it were flying in the vacuum of space, then determining the flight with aerodynamic drag included, we'll see the extreme importance of proper streamlining and shaping, because the drag-free flight will be very different from the flight with drag included. You'll see what a tremendous amount of aerodynamic force is exerted upon a model rocket in flight. Most people think that the air is quite thin and that its effects are negligible. Try telling that to somebody who's been through a hurricane! A model rocket flies faster than the winds of a hurricane.

Although model rocketry is conducted in the metric system, we'll use the more familiar American system here. Most people in the United States are still more comfortable with feet, inches, pounds, and ounces. And it's a simple matter to convert. Just refer to the conversion chart given in Table 4 of chapter 3.

To better understand what happens to a model rocket in flight, let's briefly review some of the basics about the motion of bodies in space. When a body moves from point A to point B, it covers the distance S between the two points. We'll consider only simple motion in a straight line in one dimension; add other dimensions later if you want to. Since the body cannot go from point A to point B in zero time, it takes a finite period of time T to cover the distance S.

If $S = 1$ foot and $T = 1$ second, the body is said to be moving with a *velocity* of 1 foot per second. Velocity is therefore defined as the distance traveled divided by the time required for travel. During the next interval of time, T_2, a body at constant velocity will cover another distance segment equal to S.

If the velocity of the body is not constant but changing, the body is said to be *accelerating*. It changes velocity at a given rate of change. A car accelerates if it goes from 0 to 60 miles per hour in 10 seconds, for example. If the velocity of our hypothetical body at the beginning of our observational interval of time is 1 foot per second and at the end of that interval of time it is 2 feet per second, it has changed its velocity by 1 foot per second during that period of time. If the time interval is 1 second, the body experiences an acceleration of 1 foot per second per second, or 1/ft/sec/sec, or 1 ft/sec^2.

Acceleration can change, too. But don't worry about that right now.

In 1687, the theologian and astrologer Sir Isaac Newton published a document entitled *Philosophiae Naturalis Principia Mathematica*. In this classic document, written in Latin, which was the universal language of science in those days, Newton revealed his famous three Laws of Motion. Simply stated in words these are

1. *Law of Inertia:* A body at rest will remain at rest or a body in motion will remain in motion in a straight line with constant velocity unless acted upon by an external force.
2. *Law of Acceleration:* Change in a body's motion is proportional to the magnitude of any external force acting upon it and in the exact direction of the applied force.
3. *Law of Reaction:* Every acting force is always opposed by an equal and oppositely directed reacting force.

Model rockets—and everything else in the universe that moves at speeds well below the speed of light—obey all three of these Laws of Motion. But they don't seem to make sense, do they? It doesn't seem rational to assert that if you push a body it will keep going forever and ever at the same speed and in the same direction you pushed it. Our everyday experience tells us that the body slows down and stops. But the reason it slows down and stops is the application of an external force, usually friction, acting upon the body to change its velocity in exact compliance with the Law of Acceleration.

A model rocket is one of the few accessible objects that obeys *all three* of Newton's Laws of Motion in a straightforward and easily demonstrated way. A model rocket's flight is primarily determined by the Second Law or Law of Acceleration that can be stated in the scientific shorthand of mathematics as

$$F = ma \tag{1}$$

where F = the applied force (from the rocket motor and from gravity), m = the mass of the model rocket, and a = the resulting acceleration.

If the applied force is doubled (that is, if the thrust is doubled) and the mass is kept constant (that is, the weight of the model remains unchanged), the acceleration experienced by the model will be doubled. If the force remains the same and the mass (weight) is doubled, the acceleration will be reduced by one-half.

From the basic Newtonian Laws of Motion and the basic equations of velocity and acceleration that come from them, the following general equations define the relationships between motion, distance traveled, velocity, acceleration, and time. If you're interested in how they were derived, you'll find the derivations in any high school physics text. Otherwise, just use them as they're presented here to help you determine the flight characteristics of a model rocket. Actually, these equations are greatly simplified because few of you probably know about another discovery of Sir Isaac Newton: calculus.

$$s = vt \tag{2}$$
$$V_{av} = (v_2 + v_1)/2 \tag{3}$$
$$a = (v_2 - v_1)/t \tag{4}$$
$$s = v_1 t + (at^2)/2 \tag{5}$$
$$2as = v_2^2 - v_1^2 \tag{6}$$

where s = distance, v = velocity, v_1 = velocity at the start of the time period, v_2 = velocity at the end of the time period, V_{av} = average velocity during the time period, a = acceleration, and t = length of the time period.

These equations are also very easy to program in the BASIC language of computers. This is a simple problem for a computer. Later, we'll present a BASIC computer program that takes into account *all* of the factors involved in model rocket flight, including changing thrust, weight, and air drag. Here, a computer is *really* helpful.

A model rocketeer is primarily interested in only a few things about a model rocket's flight, such as maximum altitude and coasting time. These permit a determination of the needed field size and the motor type that should be used for the flight. Maximum altitude and coasting time can be computed for the drag-free condition by a very simple method if you have three pieces of information about your model rocket:

1. The *total impulse* of the model rocket motor
2. The *burnout weight* of the model—the takeoff weight minus the propellant weight
3. The *duration* of the model rocket motor thrust

You can get the total impulse from the type code of the model rocket motor or from the manufacturer's specifications in the instruction sheet or catalog. The weight of your model rocket can be determined by weighing it on a postage scale or, more accurately, on a laboratory balance in the school science lab. Weigh it with a loaded motor and wadding installed, its exact condition at liftoff. The burnout weight can then be determined by subtracting the propellant weight (you can find the data in the instruction sheet or the manufacturer's catalog) from the maximum or gross liftoff weight you've just measured. The duration of the model rocket motor thrust is obtained from the motor manufacturer's specifications given in the instructions accompanying the motor or in the catalog.

Let's run through an example here. Suppose your model rocket has the following characteristics:

$$\text{Total impulse of motor } (I_t) = 2.50 \text{ newton-seconds}$$
$$= 0.5056 \text{ pound-second}$$

$$\text{Duration of thrust } (t_b) = 0.42 \text{ second}$$

$$\text{Liftoff weight of model } (W_o) = 39.2 \text{ grams}$$
$$= 1.38 \text{ ounces}$$
$$= 0.0864 \text{ pound}$$

$$\text{Propellant weight } (W_p) = 3.50 \text{ grams}$$
$$= 0.123 \text{ ounce}$$
$$= 0.0077 \text{ pound}$$

$$\text{Burnout weight of model } (W_b) = W_o - W_p$$
$$= 39.2 - 3.50$$
$$= 35.7 \text{ grams}$$
$$= 1.26 \text{ ounces}$$
$$= 0.0787 \text{ pound}$$

Note that we've started with all units in the metric system and have converted them. All units are in terms of pounds and seconds, including force. (One of the problems with the American system is that there is no separate unit for force as there is in the metric system.)

Recall that by definition the total impulse is the total change in the *momentum* of a body. Momentum is mass times velocity. Written as an equation,

$$I_t = m_1 v_1 - m_0 v_0 \tag{7}$$

where I_t = total impulse of motor, m_1 = mass at burnout, v_1 = velocity at burnout, m_0 = mass at liftoff, and v_0 = velocity at liftoff.

Since at zero time, or at the instant of launching, the model rocket's velocity is zero—it's sitting at rest on the launch pad—the equation reduces to

$$I_t = m_1 v_1 \tag{8}$$

When you weighed your model rocket, you actually determined the *force* with which it was being pulled toward the center of the earth by gravity. You must divide its *weight* by the acceleration of gravity (32.2 feet per second per second) to get its *mass*, which is in terms of *slugs:*

$$m = W/g \qquad (9)$$

where g = acceleration of gravity.

So equation 8 becomes:

$$I_t = (W_1 v_1)/g \qquad (10)$$

Transposing to get v_1 over to the left side of the equation by itself, we get

$$v_1 = (I_t g)/W_1 \qquad (11)$$

The term v_1 is equal to the velocity at the end of the impulse of thrust period. It is the burnout velocity of the model rocket. It is also the maximum velocity the model can attain during its powered flight and its coasting flight. Therefore, we can also call it V_{max}. And for our hypothetical experimental model rocket we can calculate

$$\begin{aligned} V_{max} &= (0.5056 \times 32.2) \div 0.0787 \\ &= 16.28 \div 0.0787 \\ &= 206.86 \text{ feet per second} \end{aligned}$$

This is the maximum velocity attained by the model rocket. It's equal to 141 miles per hour.

Remember that I said model rockets were the world's fastest models? And this example was propelled by a low-powered Type A motor!

We must now compute how high the model is at burnout (S_b). We can use equation 3 and equation 2 given earlier.

$$\begin{aligned} V_{av} &= (v_1 + v_2)/2 \\ &= (0 + 206.86) \div 2 \\ &= 103.43 \text{ feet per second} \end{aligned}$$

$$\begin{aligned} S_b &= vt = V_{av}t_b \\ &= 103.43 \times 0.42 \\ &= 43.44 \text{ feet} \end{aligned}$$

Now you know why recovery devices are not deployed at burnout of the model rocket motor! At burnout, our hypothetical model is only about 43 feet in the air and is traveling at 141 miles per hour. It hasn't achieved its maximum altitude yet, and it's moving at a speed that would tear its recovery device to shreds.

When the motor stops thrusting and the time delay starts to work, the model rocket enters the coasting phase of its flight. It coasts upward to maximum or peak altitude (apogee), trading its momentum (velocity) for altitude. During coasting flight, only gravity and aerodynamic drag act upon the model. Because we're ignoring aerodynamic drag forces in this example, only gravity is acting. The model rocket is actually falling upward in a gravity field. It's in *zero-g* or weightlessness just like an astronaut in orbit. It's being acted upon only by the acceleration of gravity, 32.2 feet per second per second. We can now go to equation 6 and find out how far upward it will coast:

$$2as = v_2^2 - v_1^2$$

Since $v_1 = 0$, then $2as = v_2^2$, and

$$s = v_2^2/(2a)$$

where s = altitude, v_2 = maximum velocity, and a = acceleration. This results in the coasting altitude (S_c) calculation

$$\begin{aligned}
S_c &= V_{max}^2/(2g) \\
&= (206.86)^2 \div 64.4 \\
&= 42{,}791.06 \div 64.4 \\
&= 664.46 \text{ feet}
\end{aligned}$$

The maximum altitude (S_t) will then be the burnout altitude (S_b) plus the coasting altitude (S_c):

$$\begin{aligned}
S_t &= S_b + S_c \\
&= 43.44 + 664.46 \\
&= 707.90 \text{ feet}
\end{aligned}$$

But what would happen if we put a Type B motor with twice the total impulse into the model rocket and flew it again? The new model parameters would be

$$\begin{aligned}
\text{Total impulse} &= 5.0 \text{ newton-seconds} \\
&= 1.12 \text{ pound-seconds}
\end{aligned}$$

$$\text{Duration of thrust} = 0.86 \text{ second}$$

$$\begin{aligned}
\text{Liftoff weight} &= 43 \text{ grams} \\
&= 1.517 \text{ ounces} \\
&= 0.095 \text{ pound}
\end{aligned}$$

$$\begin{aligned}
\text{Propellant weight} &= 6.50 \text{ grams} \\
&= 0.23 \text{ ounce} \\
&= 0.0143 \text{ pound}
\end{aligned}$$

Burnout weight = 36.5 grams
$$= 1.287 \text{ ounces}$$
$$= 0.0804 \text{ pound}$$

Running through the same calculations quickly, we find:

$$V_{max} = (1.12 \times 32.2) \div 0.0804$$
$$= 448.56 \text{ feet per second}$$

$$V_{av} = 448.56 \div 2$$
$$= 224.28 \text{ feet per second}$$

$$S_b = 224.28 \times 0.86$$
$$= 192.89 \text{ feet}$$

$$S_c = (448.56)^2 \div 64.4$$
$$= 201,206.07 \div 64.4$$
$$= 3,124.32 \text{ feet}$$

$$S_t = 192.89 + 3,124.32$$
$$= 3,317.21 \text{ feet}$$

Quite a difference! We *doubled* the total impulse, but the performance didn't simply double, did it? Although the model was slightly heavier at liftoff because of the additional propellant in the Type B motor, this heavier liftoff weight was more than offset by the motor's longer burning time. This pushed up the burnout altitude and also caused a greater burnout velocity. When we compare the two performances, the following points stand out:

1. If we *double* the total impulse, the maximum velocity increases almost *four times*—that is, the maximum velocity increases as the *square* of the increase in total impulse.
2. If we *double* the total impulse, the burnout altitude increases almost *four times*. The burnout altitude increases as the *square* of the increase in total impulse.
3. If we *double* the total impulse, the maximum altitude increases about *four times*, too.

In summary, performance increases roughly as the *square* of the increase in total impulse! This is an important relationship to remember. By running off another set of simple calculations for yourself, you will also see that any increase in weight of the model will decrease the altitude performance.

Although the numbers we've just calculated work out nicely, they don't seem to fit in with the real universe, do they? From our own experience in flying small models of this sort with Type A and Type B motors, we know that the models simply don't go to altitudes of about 700 feet with a Type A motor and more than 3,000 feet with a Type B motor. Generally, they go about 300 to 500 feet and 1,000 to 1,300 feet, respectively.

Question: What's wrong?

Answer: We deliberately ignored the effects of aerodynamic drag in these simple flight calculations.

And we can now sense the dramatic effect aerodynamic drag has on the performance of model rockets. Aerodynamic drag is a major force in model rocketry. And it's obvious that we've got to take aerodynamic drag into account when calculating flight performance.

Believe it or not, air is considered to be a fluid. And the amount of air drag experienced by a model rocket can be calculated by an equation from the science of fluid dynamics, the basic drag equation

$$D = 0.5 \rho V^2 C_d A \qquad (12)$$

where D = drag force, ρ = air density, V = velocity of the model through the air, or the air past the model, C_d = a dimensionless number called the *drag coefficient*, and A = the frontal area of the model. The drag equation tells us:

1. Air drag increases as the air density increases. How can you change the air density to reduce drag? By going to a higher altitude where there is less air density. Or by flying on a hot day, because air is less dense when it's hot.
2. Air drag increases as the *square* of the velocity. Double the velocity, and the drag force goes up *four times*. The faster the model goes, the greater the drag force becomes. See Figure 8-4.
3. Air drag increases directly as the drag coefficient increases. We'll discuss this point in detail because it's something you *can* work to reduce in model rocketry.
4. Air drag increases directly as the frontal area increases, and therefore as the *square* of the body tube diameter.

We can't do much about density except to fly at high altitudes and/or on hot days. The amount by which air drag is changed by altitude and by temperature is shown in Table 7.

The value of the drag coefficient (C_d) depends on many factors. Primarily, it depends on the shape of the model. This includes the shape of the nose; whether there are any transition pieces or changes in body diameter; the number, shape, and airfoil of the fins; the location and size of the launch lug (if any); and the smoothness of the surface finish on the model. (The size of the model is taken into account in the frontal area term in the equation.)

The drag coefficient isn't constant. It changes with change in the model's angle of attack, the angle between the long axis of the model and the direction of the air flowing past. See Figure 8-5. For most model rocket shapes, the drag coefficient increases with increasing angle of attack as shown in Figure 8-6, which was derived from model rocket tests made in the wind tunnels of the

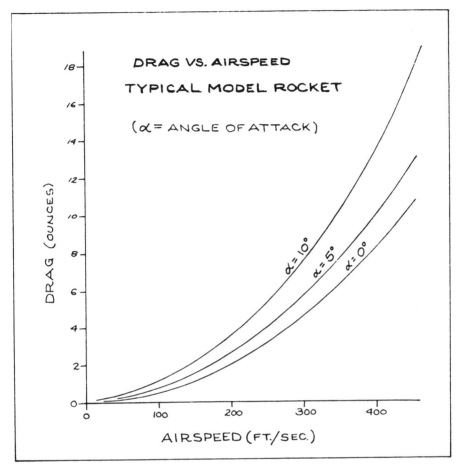

FIGURE 8-4   Drag forces versus airspeed for a model rocket at various angles of attack. Data came from wind tunnel tests at the U.S. Air Force Academy.

U.S. Air Force Academy in 1958 (data never grow old). As you can see, the frontal area of the model presented to the oncoming airflow also increases with increasing angle of attack, and this increases the value of A in the drag equation, too.

What does all this mean to you as a model rocketeer? If a model wobbles in flight, thereby flying at different angles of attack, its drag force will be greater than that of a model that slips through the air smoothly with little or no wobble. This is only one reason why you should build a stable model rocket. There are methods that can be used by a model rocketeer to alter the drag coefficient; we'll go over these in detail in the following chapter.

But how do we work the drag force equations into the flight calculations?

Answer: With great difficulty or by the use of a personal computer.

Before the advent of the personal computer, the calculations were long, tedious, and painstaking. This is because the calculations must proceed on a

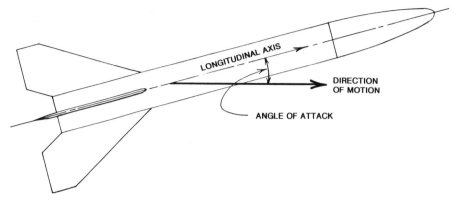

FIGURE 8-5   Graphic definition of angle of attack.

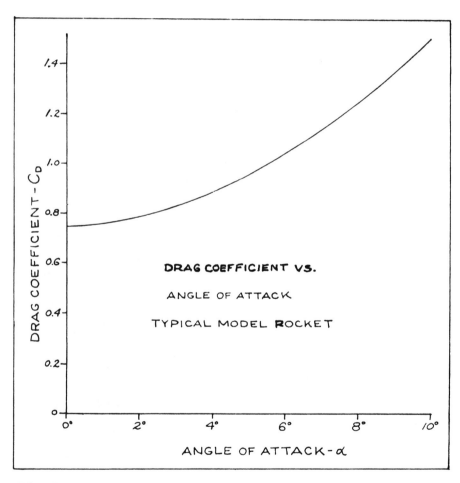

FIGURE 8-6   Drag coefficient (C_d) of a typical model rocket as a function of angle of attack.

TABLE 7
Altitude Correction Table and Temperature Correction Table

A. Altitude Correction Table
For computing change in air density as a function of elevation of launcher above sea level.

Elevation (in feet)	Multiply by
0	1.0000
1,000	.9710
2,000	.9428
3,000	.9151
4,000	.8881
5,000	.8616
6,000	.8358
7,000	.8106
8,000	.7859
9,000	.7619
10,000	.7384

B. Temperature Correction Table
For computing change in air density as a function of air temperature at launcher.

Temperature (in °F)	Multiply by
30	1.0590
35	1.0486
40	1.0380
45	1.0277
50	1.0177
55	1.0078
60	.9980
65	.9885
70	.9792
75	.9700
80	.9610
85	.9522
90	.9435
95	.9350
100	.9266

step-by-step sequential basis with the answers to one step providing the basic numbers to start calculating the next step. Also, many factors change during the flight of a model rocket.

Look at the drag equation again, and you'll immediately see that the drag force not only opposes the motion of the model through the air but also changes as the model's velocity changes. When the model takes off, its velocity is low, and the model behaves as if there were no drag. However, as the velocity begins to build up, so do the drag forces. This retards the model even more.

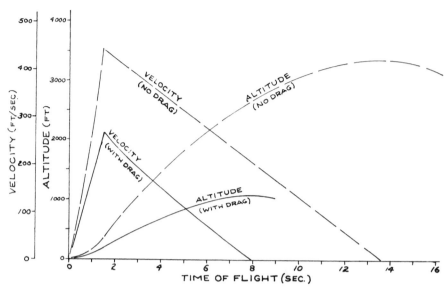

FIGURE 8-7 Comparison of computed drag-free flight and performance taking air drag into account.

Assuming a vertical flight, the acceleration of the model at any instant during its powered or coasting flight can be calculated from the equation

$$a = \frac{F - (0.5\rho V^2 C_d A)}{W - W_0} g$$

Here, a = acceleration of the model during the time interval of the calculation (usually made at intervals of 0.1 second); F = thrust of the motor at that instant (it changes during operation, as you know, and becomes zero at and following burnout); ρ = air density (which changes as the altitude increases, but for the altitudes reached by model rockets this change is so slight that we can consider air density to be constant); V = velocity of the model at the start of the time interval; C_d = drag coefficient; A = frontal area of the model; W = weight of the model at the start of the time interval; W_0 = weight loss due to propellant burning during the time interval; and g = acceleration of earth's gravity field.

I did this the long, hard way with pencil and paper *once;* that was enough. Then I got my computer and wrote a program in BASIC. The present version is called RASP-93 and is reprinted in Appendix III. Now I let my computer do these laborious calculations in a matter of minutes.

You can program your own computer, or you can download a free version of the program called pRASP, which is a port of RASP for PDA's running Palm OS. Download at www.members.shaw.ca/fc877/flightsw.html#pRasp. Although the program is copyrighted, you have permission to enter RASP-93 into your computer for noncommercial model rocket purposes. Make whatever small changes in BASIC language are required by your particular computer. It is compatible with Apple BASIC because all BASIC programs are similar. You can also modify it for use with other model rocket motors by writing a separate but similar subroutine using data from the thrust-time curve of the motor in which you're interested.

The use of modern personal computers in model rocketry has permitted us to really get a solid handle on the effects of weight, size, drag coefficient, and motor type on altitude and time of flight. RASP-93 is accurate and has been checked in the field against flights of actual models tracked as described in a later chapter. In all cases, the field data matched the computed data within the limits of motor total impulse statistical variation (plus or minus 20 percent for NAR-certified motors).

I ran the hypothetical model rocket of our example using RASP-93 in the same computer I used to write this book. I calculated for a Type A6 and a Type B6 motor with zero drag coefficient. The results were as follows:

Calculation	Motor	V_{max}	S_b	S_t
Book I_t	Type A6	208.86 ft/sec	43.44 ft	707.90 ft
RASP-93	Type A6	180.29 ft/sec	50.04 ft	546.00 ft
Book I_t	Type B6	448.56 ft/sec	192.89 ft	3,124.32 ft
RASP-93	Type B6	341.82 ft/sec	151.22 ft	1,949.37 ft

Why the discrepancies in the drag-free data? Our full total impulse mathematical calculation method assumed *constant thrust* during the entire propulsion period. On the other hand, RASP-93 uses the thrust level of the motor as determined by static testing and computes the performance in intervals of 0.1 second. The actual thrust-time curve of the Type A6 motor is not steady but more like the spike of a core burner. This real-world thrust performance results in a higher burnout altitude but a lower maximum velocity. In contrast, the Type B6 motor has a high initial spike but a long and lower constant sustainer thrust; the combination of the spike plus low sustainer thrust gives us much lower altitude for the drag-free computer calculation than for the constant-thrust hand-calculated method.

The RASP-93 BASIC computer program handles the change in thrust, change in weight, and change in velocity of the real world of model rocket flight. Because it computes the entire flight in time increments of 0.1 second, it's much closer to reality than programs that average various values. Using RASP-93, I recalculated the flight of our hypothetical model rocket with a drag coefficient included. I chose a drag coefficient of 0.75, which seems to be about average for most sporting model rockets. The body tube was 0.984 inch in diameter. So our hypothetical model rocket is a small, inexpensive, single-stage sporting bird similar to an Estes Alpha or a Quest Astra. Let's compare the digested results of the RASP-93 drag-free ($C_d = 0$) and medium-drag ($C_d = 0.75$) cases:

Motor	C_d	V_{max}	S_b	S_t
A6-5	0	180.29	50.04	546.00
A6-5	0.75	173.09	48.99	306.99
B6-5	0	341.82	151.22	1,949.37
B6-5	0.75	300.70	141.24	611.29

These numbers are interesting and lead to some fascinating conclusions. First, air isn't as tenuous as we might think because it certainly seems to have a profound effect on the flight of something as small and streamlined as a model rocket. Second, the computer calculations with a reasonable drag coefficient included seem to be in the ballpark, because we know that a Type A motor will take a model to about 300 to 500 feet and a Type B will lift it up to between 600 and 1,000 feet.

Of interest as well is the fact that these modern computer analyses have caused us to change some of our previous concepts about model rocket performance. Now it appears that we can draw the following conclusions about model rocket flight performance:

1. The aerodynamic drag of a model rocket has a very small effect on the burnout altitude of a model rocket unless high-powered motors are used, which will in turn increase burnout velocity and thereby increase air drag and lower the drag-free burnout altitude. Aerodynamic drag *will*, however, decrease the burnout altitude if the drag coefficient or frontal area is large.
2. Aerodynamic drag lowers the computed drag-free maximum altitude of a model rocket by about 50 percent (for low-powered models) to as much as 80 percent (for high-powered models).
3. Aerodynamic drag forces on a model rocket become very large at velocities of 100 feet per second or more, requiring very rugged construction for models designed for high performance and propulsion by Type D, Type E, Type F, and Type G motors.
4. The highest drag forces and the greatest structural stresses on a model rocket occur at or near burnout and rapidly become less as the model coasts up to apogee. In our hypothetical model, *negative acceleration* of 2.7 g's would be experienced due to air drag within a tenth of a second after burnout and rapidly drop to 1.5 g's a second later.

Another interesting fact is hidden in all these equations and calculations. You can use RASP-93 to discover it for yourself. The drag-free calculations would lead you to believe that the lighter the model rocket, the higher it will go. This just isn't true when you bring aerodynamic drag into the picture. In the drag equation, both the drag coefficient and the frontal area are divided by weight. Therefore, it's perfectly possible (and I've built model rockets that prove it) to reach a point where an ultralight model rocket acts just like a feather—and you can't throw a feather very far! The model's area-to-weight ratio or drag-to-weight ratio gets to be so large that the aerodynamic forces completely overwhelm the momentum (mass times velocity) forces. The bird staggers up to 30 feet with a Type C6 motor in it and stops dead at burnout.

This gave birth to the competition event known as drag racing, and it's just what the name implies. The two-model heat goes to the bird that scores two out of three points: (1) getting off the pad first; (2) going to the *lowest* altitude of the pair; and (3) touching the ground *last*.

In contrast, if you make a model rocket too heavy, the thrust from the motor will be unable to accelerate it to a high velocity, and the peak altitude will be low. Most of the model rockets built for Type A through Type C motors are too heavy. Most of the higher-power model rockets are too light.

Therefore, a model rocket designer gives a little here, takes a little there, trades this for that, and tries to reach a happy compromise. For some reason, the early model rocket designs were right at or very near the optimum trade-off between weight and drag. This was a fortunate circumstance because it permitted the historic models to operate quite satisfactorily, giving model rocketry time to develop the nuts and bolts of its technology before having to get involved with technical trade-offs that we can now determine with modern personal computers.

Analysis of flight dynamics and trajectories is a favorite subject for college undergraduates who've been model rocketeers and have taken their hobby to college with them. The Massachusetts Institute of Technology (MIT) Model Rocket Society has done a lot of work in this area. Cadets at the U.S. Air Force Academy have utilized their wind tunnels for further drag studies. Other work has been done at Ohio State University, the University of Maryland, and Kent State University by model rocketeers who've decided to become professionals. Yet this is an area where junior high school students can work with minimal mathematical tools and abilities and where students can learn to program and use computers.

It's possible today to predict with exceedingly accurate results the altitudes, accelerations, and velocities of model rockets. These numbers are quite helpful in designing model rockets properly. And it's very exciting to see the model rocket you've designed and flown in the computer perform in the real world just as you calculated it would. It's an example of how engineers and scientists can accurately predict the future by knowing how the universe works and how to make it work for them—and for the rest of us, too.

STABILITY

9

In the preceding chapter, we saw that the earth's atmosphere—the air—plays a major role in the flight of a model rocket. Aerodynamic drag greatly reduces performance. But by utilizing careful design techniques based on a century or more of scientific research in aerodynamics, we can *use* the air to stabilize the model, keep it going in the intended direction, and make it fly predictably. We can also decrease the aerodynamic drag and increase the performance of a model rocket by understanding how the air flows around it and creates the drag force.

We must always keep in mind that a model rocket is a free body in space after it leaves the launch rod. It's not attached to the ground in flight, and the forces acting upon it in flight cannot be easily duplicated on the ground.

In discussing the performance of a model rocket in flight, we acted as observers on the ground, watching the model with reference to the ground. We were standing still and watching the model move. To better understand aerodynamic stability and aerodynamic shapes, we must change our point of view and travel with the model in its flight through the air—that is, we must become theoretical passengers in order to see and understand better what is happening.

A model rocket in flight can move in *eight* different ways. This is technically known as *eight degrees of freedom.* For simplicity and ease in considering the motions, we can reduce any motion of a model rocket in flight to a combination of one or more of the eight basic motions shown in Figure 9-2.

Thrust moves the model rocket forward. It comes from the model rocket motor. *Drag* opposes the thrust force and attempts to slow the model. (The *gravity* force can come from any direction, depending on the altitude and flight direction of the model. We've already taken the gravity force into account in calculating the flight dynamics of the last chapter.)

Yaw is a swinging motion of the nose to left or right. *Pitch,* an up or down motion of the nose, is similar to the yaw motions. Because a simple model

FIGURE 9-1   The size and shape of model rockets determine the air drag they will encounter in flight. Some shapes have less drag than others.

rocket is the same shape in both the pitch and yaw aspects, the two motions are usually lumped together in the term *pitch motion*. (But this isn't true for most boost gliders and rocket gliders, as we shall see in a later chapter devoted to them.) *Roll* is a rotational motion where the model spins right or left about its long axis.

Thrust, drag, and gravity forces are *linear;* they produce motions in a straight line that are called *translational motions*. Pitch, yaw, and roll are *rotational motions*.

Unlike a body experiencing a translational motion, anything that rotates must have a rotational axis, an imaginary line around which it spins. The earth, for example, has a rotational axis running through the North and South Poles. A model rocket has a roll axis that is an imaginary line running through the tip of the nose down along the center line of the model and out the nozzle

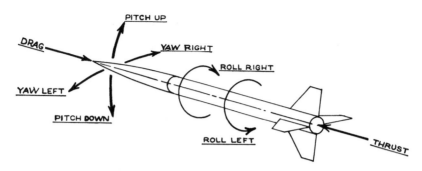

FIGURE 9-2   The eight different degrees of freedom of a model rocket in flight. Only two—thrust and drag—are linear. The other six are rotational.

of the motor. There is also a pitch axis and a yaw axis, imaginary lines through some points in the model.

Where are these rotational points or centers of rotation? How do we find them?

When a free body in space rotates, it spins around an imaginary point where all its mass seems to be concentrated. This is its balance point. It is called the *center of gravity,* or CG.

You can perform a simple experiment to prove that a body rotates around a single point that is also its CG. Take a stick, a wooden dowel, or a body tube about 18 inches long. Balance it carefully and mark the balance point with a felt-tip pen, making a line all around the stick or tube at that point. Toss the stick or tube into the air with an end-over-end motion. You will easily see that the stick or tube spins around the balance point you've marked. No matter how you throw the stick or tube to get an end-over-end motion, it will always rotate around this point.

Now put some putty or plasticine clay firmly on one end of the stick. Rebalance the stick and mark the new CG point, using a different color felt-tip pen. The CG will be in a new location, closer to the end on which you put the additional weight. Again throw the stick end over end and you'll see that it now rotates around the new balance point or CG.

A model rocket in flight will rotate around its CG in the pitch, yaw, and roll axes. Why be concerned about these motions? Because if a model rocket rotates around its pitch or yaw axis, it's going to change its direction of flight. And you want the bird to go right where you pointed the launch rod: straight up. If it leaves the launch rod only to spin around its pitch and yaw axes, its angle of attack and therefore its aerodynamic drag will increase, and the model will also go in a direction other than the way the launch rod is pointed. These other directions can often be very erratic and unpredictable, to say the least.

What can cause a model to rotate around its pitch-yaw axis? As Sir Isaac Newton said in a voice that calls down over the centuries to the rocketeers of today: *A change of motion of a body can be produced only by an external force acting upon that body.*

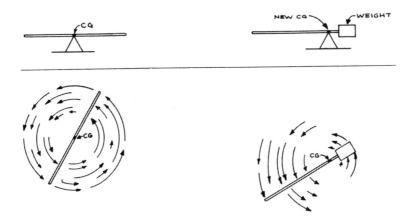

FIGURE 9-3   Rotation of a body about its balance point or center of gravity (CG).

There are many external forces that can cause rotational motions in a model rocket—gusts of wind blowing at the instant of launch, winds blowing at various altitudes during flight, fins crookedly positioned on the body tube, off-center thrust from the motor, irregularities in model construction, and many others. No matter how perfectly you build a model and no matter how carefully you try to launch it under ideal conditions, there will always be some tiny forces that will begin to produce pitch-yaw rotational motion the instant the model leaves the launch rod. You cannot eliminate them.

Therefore, the model must incorporate some sort of stabilization device that will overcome these motions, damp them out, and restore the model to its intended flight direction and altitude. Furthermore, this must be done in a fraction of a second. A force must be almost instantly created to oppose the rotational force.

Most space rockets counteract rotational forces with an automatic electronic control system, an autopilot that uses gyros to sense the rotation in the pitch-yaw axis, then sends electrical signals to a computer, which in turn signals various hydraulic devices to tilt rocket motor nozzles or to fire attitude-control rocket motors to correct the effects of the disturbing force. These control systems are, in comparison to model rocket technology, heavy, large, complicated, and expensive. Although some model rocketeers have experimented with tiny gyros for controlling their birds, most model rockets get their stabilizing and restoring forces in a simple manner from the air rushing past the model and acting upon aerodynamic surfaces—the fins.

When a moving stream of air strikes a surface broadside (at an angle of attack of 90 degrees) or even at a slight angle, it produces a high pressure on one side of the surface and a low pressure on the other, as shown in Figure 9-4. This pressure difference creates a drag force opposite to the motion of the airstream and a lift force that is at right angles (90 degrees) to the surface. The higher the angle of attack, the greater the lift and drag forces—up to the angle

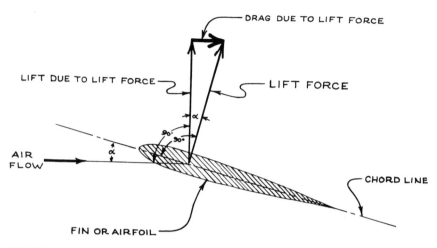

FIGURE 9-4   Lift force and drag force due to lift of a fin (shown in cross-section) at an angle of attack. Lift force is always at right angles to the chord line.

of attack where the surface stalls. At the stall point, the air breaks away from the low-pressure surface, the lift force decreases drastically, and the drag force increases tremendously. By properly positioning the fins on the model rocket, by making them the right size, and by giving them the right shape, we can use this lift-drag force as a stabilizing force to offset pitch-yaw rotational disturbances.

As we saw with our tumbling stick, we can make the simplifying assumption that all the mass (weight) of a body is concentrated at its CG. It certainly acts this way, doesn't it? Let's extend this concept to *any* force acting upon the body, including the aerodynamic lift-drag force we've just introduced. Therefore, we can think of the body as having a point where all air pressure forces act. We call this point the *center of pressure,* or simply the CP.

The function of the CP can be illustrated by another experiment with the stick or tube. If we grasp the stick at its balance point (CG) between a pair of pointed, low-friction pivots as shown in Figure 9-5, and hold it in the moving air from an air-conditioning duct (*not* from a fan because of the turbulence generated by a fan) or out of the window of a car traveling at about 20 miles per hour, we'll see that the stick doesn't rotate in the pivots due to any air pressure forces. No matter how we hold the stick in the airstream, it doesn't rotate. Obviously, there are no off-center or unbalanced forces to make it rotate, so the CG and the CP must be the same. We'd expect this because of the symmetrical shape of the stick.

Now glue or staple a piece of stiff cardboard to one end of the stick. Because of the additional weight of the cardboard vane, rebalance the assembly to find the new CG. Pick up the stick-plus-vane with the pivots at the new CG and place it in the moving air again. The stick will immediately swing

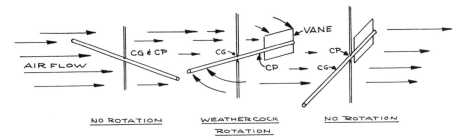

FIGURE 9-5   The stick-and-vane model under various conditions described in the text.

around with the vane downwind and the stick pointed directly into the wind. Congratulations! You've just reinvented the weathervane.

The presence of the cardboard vane produces more air pressure force on its end of the stick. This air pressure force was caused by the lift-drag of the vane. The presence of the vane also put the CP at a different point than the CG. There's now a difference between the points at which two different forces act.

If you push the stick slightly to the side with your finger, you'll discover that the air pressure force becomes greater as you displace the stick more and more from the nose-into-wind position it naturally seeks to maintain. There is very little force when the angle of attack is low, but more force is created as the angle of attack increases. This is the sort of restoring force that will stabilize model rockets.

If we start to move the pivot point closer to the vane end of the stick, we'll eventually find the point where the stick will no longer pivot into the wind. This is the point where the air pressure forces on both ends of the stick are equal. We've just located the CP of our stick-vane model.

We can replace the stick and vane with a model rocket and it will behave in the same manner. When picked up and supported by pivots at its CG, the model will always point into the airstream. With both models, the CP is *behind* the CG, making it an aerodynamically stable situation.

What would happen if we balanced either of our models so that the balance point, the CG, was behind the CP, then picked it up with the pivots at the new rearward CG location? Very interesting! The model would try to fly tail-first. A real model rocket under thrust doesn't want to fly tail-first because of the thrust force. It becomes wildly unpredictable in flight, thrashing all over the sky because it's an unstable model.

As shown in Figure 9-6, there are basically three stability conditions for any body, including a model rocket—and only one of them is desirable for a model rocket. They are

1. *Positive stability,* where in a model rocket the CG is ahead of the CP. It has large fins set far back on the body tube. It will fly straight when launched and will weathercock into the wind at launch.

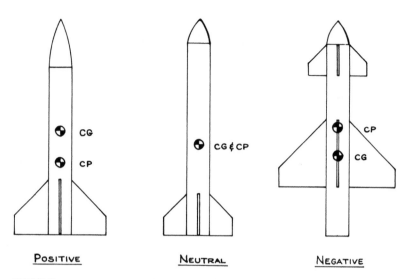

FIGURE 9-6 The three stability conditions with their CG-CP relationships.

2. *Neutral stability,* where the CP and CG lie at the same location on the model. This might be caused by a lightweight nose or by fins that are too small, or both. There are no stabilizing and restoring forces present in the model during flight. It's free to wander anywhere in the sky, and some of its wanderings may be wild and certainly unpredictable. It may become stable or unstable at any moment because of the burnoff of the propellant, then it might keep right on going in the direction it happens to be pointed at that instant.

3. *Negative stability,* where the CG lies behind the CP. In this case, the aerodynamic forces on the fins try to make the model fly tail-first, which it doesn't want to do under power. Once the nose swings in pitch or yaw after leaving the launch rod, a force exists to keep it swinging. The unstable model usually pinwheels end over end and winds up going nowhere except to flop to the ground.

Remember the positive stability condition by the mnemonic or memory aid: *G before P.* This is the alphabetical stability rule, because the CG comes before the CP in a stable, flying model rocket.

You should *never* fly a model rocket other than kit designs until you have determined its CG-CP relationship to ensure that it will be stable in flight. The easiest way to make this determination is by the swing test. Tie a 6-foot length of string around the body tube of the model at the balance point. The model with a loaded motor installed should hang in a horizontal attitude from the string—that is, the string is around the model at its CG. Hold the string in place with a bit of tape so that the string doesn't slip.

FIGURE 9-7 The easiest way to determine the stability of a model rocket is to conduct a swing test.

Then, making sure that nobody's in the way, start to swing the model around your head in a horizontal circle. The longer the string, the more valid this test. If the model is stable, its nose will point in the direction you're swinging it. If the nose points elsewhere, you must add some weight to the nose to bring the CG forward.

Don't be dismayed when I suddenly reveal to you here that the CP of a model rocket usually depends on its angle of attack. The model may exhibit good stability if it swings to a small angle of attack. But if it is displaced to a large angle of attack, or if a strong gust of wind hits it in real flight conditions, it may not be able to recover its stable condition. This is because the CP of most model rocket designs moves forward as the angle of attack increases. It reaches its most forward point when the angle of attack is 90 degrees. Of course, some model rocket designs have less CP movement than others.

How does a model rocket designer handle this mess? Quite easily.

In the early days of model rocketry, we determined the CP location by letting it equal the center of lateral area. To determine this point, we made a cardboard cutout of the side-view shape of the model. When we balanced this cardboard cutout, it gave us the center of lateral area, which we used as the CP. Actually, the center of lateral area is indeed the CP of a model rocket if it's flying at an angle of attack of 90 degrees, the worst possible condition.

All the models designed with this cardboard cutout method of locating CP flew successfully. But there were a few models that just "happened" and

FIGURE 9-8   Checking the stability of a model rocket using a cardboard cutout of a side view of the model, thus locating its center of lateral area.

weren't really designed. They flew well even though the cardboard cutout method of CP location indicated they wouldn't be stable.

The research and development department at Quest took the cardboard cutout idea one step further to actually flight test three fin designs. You can create a three-dimensional cardboard model by cutting out two identical images of your design, scoring them down the middle, then folding along the score so that you have two mirror images of the rocket profile. Then use a glue stick to glue the two halves together. Using the same method of balancing the model on the edge of a ruler, you can determine where the CG is. Normally it starts out as neutral (same position as the CP). You then add small amounts of modeling clay to the nose and hand toss the model several times until it begins to fly in a stable manner. This method is a quick and inexpensive way to test your designs.

Model rocketeers knew from NASA technical reports that aeronautical engineers and fluid dynamicists had a method of computing CP. However, it was very complicated and was used to determine the supersonic CP of rocket shapes. Nobody seemed to be able to apply it to model rockets flying at subsonic speeds.

The breakthrough came in 1966 when James S. Barrowman, a professional aerodynamic engineer with the Sounding Rocket Branch of NASA's Goddard Space Flight Center, who was a model rocketeer and also became president of the NAR later, presented a simplified technique for computing the actual CP of a subsonic model rocket at low angles of attack.

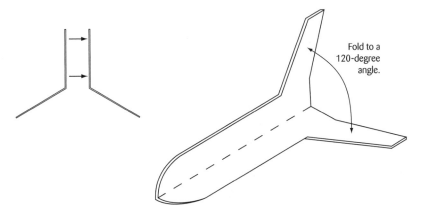

FIGURE 9-9 A hand-toss flight test model can be made from two mirror-image cardboard cutouts of your rocket.

The Barrowman method was extremely successful. I've even used it to calculate the CP of full-size rocket vehicles for some studies on launching dynamics and how to ensure that the rocket remained stable through the subsonic phase of flight.

The equations and procedure of the Barrowman method are shown in Appendix II. These may seem complicated at first, but a lot of model rocketeers have mastered the Barrowman method.

Thanks to modern personal computers capable of being programmed in BASIC, you can do design work on very complicated model rockets. I've written a BASIC program, STABCAL-2, for doing this. It will handle a model rocket with up to three transitions and up to three sets of fins. Thus, it will handle the stability calculations for three-stage models as well as simple ones. For simple models, the program doesn't bother to calculate transitions or fins you indicate aren't there. STABCAL-2 is presented in Appendix IV; you may have to modify it slightly to conform to the conventions of the particular form of BASIC your computer uses, but these shouldn't be extensive. There's only one square root calculation involved and the rest is simply arithmetic operation. A commercially available software program called SpaceCad can also be used. You can download a free demo at www.spacecad.com.

If you don't want to bother with any of this, you can get along fine by using the good old cardboard cutout method. Your models will have too much fin area and will have a tendency to weathercock more in a wind, but they will fly in a stable manner.

THE QUESTION OF STABILITY

How much stability should a model rocket have? How far behind the CG should the CP be located?

It is generally agreed among advanced model rocketeers and has been generally confirmed by flight tests that the CP should be no less than one body diameter behind the CG. If the body tube diameter is 1.34 inches, the CP should be at least 1.34 inches behind the CG. This is known as *1-caliber stability*. In gunnery terms, *caliber* refers to body diameter, and the word comes to us from the days when rockets were part of the artillery corps of armies.

Technically, one-caliber stability is all you really need for most sporting models. It allows for the rearward movement of the CP with increasing angle of attack up to a reasonable limit beyond which your model probably won't go, anyway. If your model has more than 2 to 3 calibers stability, it may be overstable and suffer from excessive weathercocking (which may be something you want in a parachute duration type of contest model). It will certainly have fins that are too big, adding additional unneeded weight.

What do you do if your model rocket checks out as neutrally stable or negatively stable—or if it is less than one-caliber stable? That depends on the model. It may be possible to increase the fin area or to move the fins back; this moves the CP rearward. Such an approach may not be possible with a scale model where the size and location of the fins are predetermined by the dimensions of the real, full-size rocket. In the case where you can't change the size, shape, or location of the fins, you must add weight to the nose of the model to bring the CG forward. Sometimes you must do a little bit of both to obtain the optimum weight and the least amount of fin area. This is one of the things that makes a model rocket designer's work so fascinating. There are so many trade-offs that can be made.

For many years—and even now in some parts of the world where model rocketry, or space modeling, is new—model rocketeers believed that they should make the smallest, lightest model rockets possible in order to achieve maximum performance. As we've seen in the previous chapter, there's a serious flaw in this logic because it's possible to make a model rocket so lightweight that it tries to fly like a feather. This same make-it-smallest philosophy also resulted in some short, squat, fat little model rockets that were barely large enough to enclose a motor, a very small recovery streamer, and a nose cone. To obtain the proper stability, these models had fins that were sharply swept back behind the model. True, they had one-caliber stability. But all of the major weights of the model—motor, nose, and recovery device—were located very close together.

In flight, these models wobbled back and forth very rapidly as they ascended. We found out earlier that drag increases as the angle of attack increases. Therefore, these short squatty models experienced substantial drag because of their constantly and rapidly changing angle of attack. On the other hand, many of us had started to fly long, slender models that weighed more but would slither upward with little or no wobble. The basic performances are shown in Figure 9-10.

To understand why the two designs functioned as they did, let's go back to our stick model held between low-friction pivots out of a car window, our moving wind tunnel. Make a short stick about an inch in diameter and about

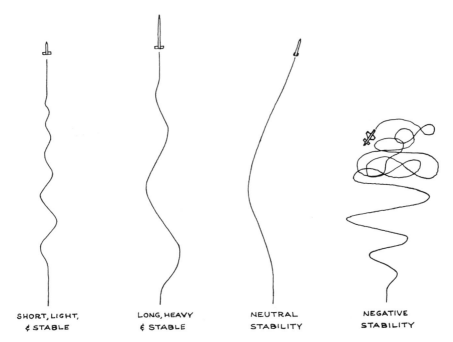

SHORT, LIGHT,
& STABLE

LONG, HEAVY
& STABLE

NEUTRAL
STABILITY

NEGATIVE
STABILITY

FIGURE 9-10 Flight paths of model rockets with various static and dynamic stability characteristics.

4 inches long. Put a cardboard vane 2 inches by 4 inches on it. Make a longer stick about an inch in diameter and about 12 inches long. Cut a cardboard vane 1 inch by 2 inches and staple it to the stick (see Figure 9-11).

Grasp the short stick model by the pivots at the CG and put it into the airstream. Displace it in the yaw direction with your finger and watch how it behaves. It's a stable model. The air hits the vane and produces a restoring force. The stick starts to swing back to zero angle of attack. It does so very quickly because all its mass is concentrated near its CG. Very little inertia is involved. The stick will reach its maximum rotational velocity when the angle of attack is zero and it will not stop swinging. It will swing through zero angle of attack and take up an angle of attack on the other side of zero. The stabilizing situation is reversed and the model then starts to swing back again. It will oscillate, swinging back and forth on either side of the zero angle of attack point. The oscillation will be very rapid, and several swings will be required before they damp out and the model stops swinging.

Now hold the long stick with the small vane in the airstream in the same manner. It's also stable. Displace it in the yaw direction as you did with the short model. When released, the long model will also start swinging back, but it will take longer to do so. Its rotational velocity will be much less as it swings through zero angle of attack. It will also swing to the other side. But its oscillations

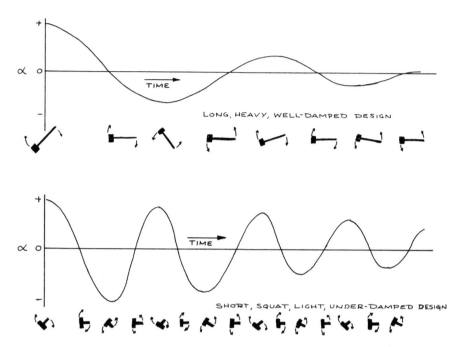

FIGURE 9-11   Visualization of dynamic stability with the stick-and-vane models described in the text.

through the zero point will be slow, will be fewer, and will damp out after perhaps only one or two passes through zero angle of attack.

These actions are caused by very complex phenomena grouped together under the general classification of *dynamic stability.* Just because we can hang a name on it doesn't mean that we understand it any better. Basically, the difference between static stability and dynamic stability is that of balancing a nonmoving device versus a moving one.

It is perfectly possible for a model rocket to have its CG and CP in the proper locations for good static stability while actually being dynamically unstable in flight. Model rockets encounter several kinds of dynamic stability problems:

1. *A statically stable but dynamically undamped model.* This is represented by the short, squat, fat little model that wobbles excessively as it flies.

2. *A statically stable but dynamically unstable model.* In this model, the CP and CG may be properly located with respect to one another, but the model is too heavy and has fins that are too small. These small fins, while creating a statically stable model with a CP behind the CG, are too small to produce enough restoring force to return the model quickly to zero angle of attack. They aren't large enough to

overcome the turning momentum of the model. By the time these small fins can create sufficient restoring force, the model is doing something else.

3. *A statically stable but dynamically overdamped model.* In this condition, the model might have a long, skinny body with fins that are far too large. When the model weathercocks or rotates in the pitch-yaw axes, the fins produce too much restoring force and stabilize the model too quickly, causing it to fly as if it were almost neutrally stable. This condition is rarely seen, but it can exist with some canard-type boost gliders where the wings are large and at the rear of the model.

4. *A model with pitch-roll coupling.* This is a weird form of instability that can really frustrate you if you don't know about it. In this situation, the model has some roll that's induced by fin, nose, or motor misalignments. At some point in the flight, the frequency of the roll becomes the same as the frequency of the motion back and forth in the pitch-yaw axes. The model will start to exhibit a coning motion where it spins about the roll axis at the same time that it begins to rotate in both pitch and yaw about the CG. The model can become completely unstable as it rolls madly around its long axis and spins horizontally end over end, going nowhere. The problem of pitch-roll coupling occurs in full-scale fin-stabilized rockets.

These dynamic stability problems were first mentioned in an earlier edition of this book. We really didn't know very much about them then. They were complex and were not yet actually understood by professional rocket engineers, either! There was a lot of discussion among rocketeers, professional and model, but nobody really had a handle on *any* of the dynamic stability problems. (If you think these are bad for simple rockets, they're life-and-death matters for people who ride in asymmetrical airplanes that suffer dynamic stability problems involving such things as flutter and yaw damping!)

As a result, a model rocketeer named Gordon K. Mandell tackled the subject while he was still an undergraduate at MIT. He improved the MIT low-speed, low-turbulence wind tunnel and developed some ingenious instruments to measure motion and forces on model rockets placed in that wind tunnel. He also made some simplifying mathematical assumptions and succeeded in linearizing the equations to the point where they could be useful in producing results for model rocket designers.

Mandell discovered that most of our model rocket designs were *overdesigned* on the safe side. This wasn't surprising, since most of us tend to be a bit conservative in design because of all the unknown factors that we deal with and all the unforeseen conditions of a model rocket flight. No real rules of thumb have yet been developed from Mandell's classic work. Perhaps you or some other model rocketeer will pick up where Mandell left off so that these rules of thumb can be reported in a future edition of this book. Briefly, Mandell stated some basic points of design to keep in mind:

1. Maintain a length-to-diameter ratio of 10:1 or more to provide adequate damping.
2. Maintain a static stability margin between 1 and 2 calibers to prevent overdamping, but don't go below 1 caliber.
3. Hold the roll rate of the model as low as possible to prevent pitch-roll coupling, and align the fins carefully in an attempt to get zero roll rate for best performance.
4. If you must increase the linear dimensions of the fins to get proper static stability, increase the *span* dimension (the dimension outward at right angles to the body tube), because this will increase the restoring force rather than increasing the distance of the CP from the CG, which in turn improves the dynamic damping.

We now have static stability well in hand and understood, and Mandell has made significant progress toward solutions of the dynamic stability problems we run into from time to time. But there's still a lot of work to be done by model rocketeers who get interested in this area of the hobby and who want their work to be recognized.

This is one reason why model rocketry should never be considered just kids' stuff. Barrowman's work and Mandell's findings have been used in design work on real sounding rockets. Interestingly, Harry used these principles to help in the design of a space rocket when the engineers didn't know the subsonic stability factors because they'd handled only supersonic rockets during their careers. Harry did a swing test on a scale model but didn't tell the engineers he'd done it that way until after they had confirmed his calculations. When he did tell them, they didn't believe him until he demonstrated it to them with the model! Professional rocket engineers at NASA have used Mandell's work because it's just as applicable to full-size rockets as to our models. It took a model rocketeer to get the complexities of dynamic stability simplified to a workable point.

10

MODEL ROCKET AERODYNAMICS

Aerodynamics involves the way air flows around various shapes and the forces that are created in this process. Thus far, we've been discussing some areas of aerodynamics without delving deeply into the subject itself. But, as we've discovered, aerodynamics plays a large role in model rocketry. Model rockets are stabilized by aerodynamic means. And air drag has a major effect on model rocket flight.

Aerodynamics is a branch of the science of fluid dynamics. However, it's concerned with only one fairly complex, composite gas made up of oxygen, nitrogen, carbon dioxide, and a few other elements. This special gas is the earth's planetary atmosphere. Aerodynamics may someday be extended to other planets that have atmospheres such as Venus, Mars, Jupiter, Saturn, Uranus, and Neptune.

Many years ago, inventors and "aeroplane" builders learned that some shapes have less drag and create more lifting force than others. A tremendous amount of information is available on the subject of aerodynamics, some of it going back more than a century. For example, some of the data used by James S. Barrowman in developing his simplified CP calculations came from research done during World War I for biplanes such as the Sopwith Camel and the SPAD. Thus, our Space Age hobby of model rocketry has technical roots that go back to the days of the Red Baron and Eddie Rickenbacker. All of these data are readily available to model rocketeers in textbooks and government reports. Many of them have used the data.

Moving air can exert considerable force on an object, as we've seen in the previous discussions of flight dynamics and stability. Uncontrolled, aerodynamic force can tear a model to pieces. Properly used, it can stabilize a model rocket. Much of the force generated by air in motion depends on the shape of the body around which the air flows. If you put your hand out the window of a car at 55 miles per hour (carefully, please) and open your palm broadside to the airflow (that is, at a 90-degree angle of attack), you'll feel a definite push

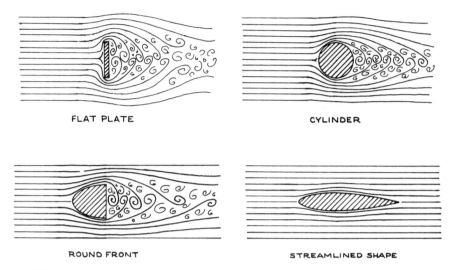

FLAT PLATE   CYLINDER

ROUND FRONT   STREAMLINED SHAPE

FIGURE 10-1   Visualization of typical airflow patterns around objects of various shapes.

against your palm. If you make your hand into a fist, this semispherical shape will have less aerodynamic drag, and you'll feel less push from the air.

Basically, the drag force is caused by the need for the object to push the air molecules out of the way, slide through them, and permit them to close in behind it again with the least amount of disturbance. Figure 10-1 shows how air can be visualized as flowing around some objects of different shapes. Thus, the shape of an object and the pattern of airflow around it have a major effect on the amount of drag force produced.

Several forms of air drag are of interest to model rocketeers. They may be summarized as follows:

1. Friction drag
2. Pressure drag
3. Interference drag
4. Parasite drag
5. Induced drag

Let's explore each of these in turn, because drag is very important to model rocket flight. Air is made up of molecules and is a mixture of gases. For our purposes in model rocketry, the molecules can be considered homogeneous air molecules rather than a mixture of different gas molecules. These air molecules are so small that trillions of them would fit on the period at the end of this sentence. You can think of multitudes of molecules as tiny Ping-Pong balls separated from each other by very small distances. At the earth's surface under normal conditions of air pressure and temperature, the average air molecule can travel only 0.000002419 inch (29-millionths of an inch) before hit-

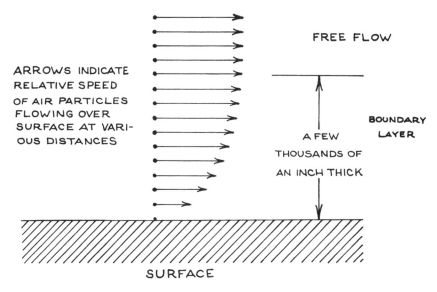

FIGURE 10-2   The boundary layer.

ting another air molecule. So the earth's atmosphere is a crowded place for air molecules. The Ping-Pong balls hit one another constantly, creating a gross effect we call *pressure*.

When the molecules slide over the surface of a nose, body tube, or fin, friction results between the surface and the molecules and between the molecules themselves. Actually, although the surface may seem to be mirror-smooth to the eye, it's a mass of microscopic hills and valleys. *Friction drag* is caused by the air molecules bumping into hills, rebounding off valley walls, and bumping into one another as a result. The rougher the surface, the more numerous are the microscopic hills and valleys for the molecules to hit, and the greater the friction drag.

Actually, the airflow next to the surface exhibits some rather strange and unexpected activity. Right on the surface, the friction and the viscosity (internal friction) of the air slow the first layer of molecules almost to a standstill. The next layer slips and slides over the first layer at a little higher speed. So it goes, with each successive layer sliding faster over the layer below until the full free-stream velocity is reached. This fluid flow phenomenon that takes place close to a surface is called the *boundary layer*. It's shown diagrammatically in Figure 10-2.

All objects moving with respect to the air create a boundary layer. The thickness of the boundary layer varies with the size of the model and the speed of the airflow. If the layers are slipping over one another in an orderly fashion as shown in Figure 10-2, the average model rocket traveling at an airspeed of 250 feet per second will have a boundary layer only about $1/1,000$ inch thick.

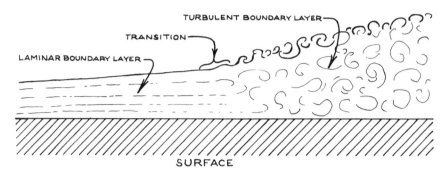

FIGURE 10-3   Boundary layer transition from laminar to turbulent state.

You can experience a boundary layer on a beach. If you lie down on the sand, you'll be in the boundary layer of the beach surface. You'll feel less wind speed there than when you're standing.

The boundary layer can be *laminar* as shown in Figure 10-2, or it can be *turbulent* as shown in Figure 10-3. When the boundary layer undergoes the *transition* from laminar to turbulent, the layers of air molecules in the boundary layer no longer slip easily over one another. They swirl and eddy about within the boundary layer itself, even though the boundary layer is still attached to and flowing along the surface.

You can see laminar and turbulent boundary layers very easily by playing with the faucet in the kitchen or bathroom sink, unless the faucet has an aerator. If it does, take off the aerator. Open the faucet slowly and carefully until the water streams out in a smooth, clear fashion. Then carefully open the faucet a little bit more. The stream will suddenly break into turbulence a short distance below the spout. If you continue to open the faucet and increase the flow, the entire stream will become turbulent. Like air, water is a fluid and obeys the same rules of fluid dynamics even though it has a higher density and will not decrease its volume under pressure as air does. (Water is known as an *incompressible fluid,* whereas air at subsonic velocities is a *compressible fluid.*)

With even the smoothest surface, the boundary layer becomes turbulent at some distance along a model rocket. This distance is a function of the size, shape, smoothness, and speed of the model. Because of the swirling eddies of a turbulent boundary layer, friction drag is much higher than it is in a laminar boundary layer.

A very small irregularity on the surface will cause transition from laminar to turbulent boundary layer conditions. For most model rockets flying at 250 feet per second, a protuberance 0.0003 inch high will cause transition or trip the boundary layer from laminar to turbulent.

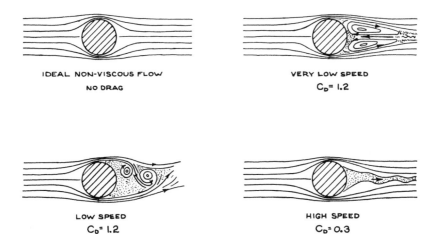

FIGURE 10-4   Visualized airflow around a cylinder under various conditions.

Data indicate that the boundary layer on a model rocket usually trips on the surface of the nose usually very close to the nose base. Some modelers theorize that it's best to deliberately trip the boundary layer at the nose-body joint because it's going to trip there, anyway. Deliberately tripping the boundary layer means that it will be tripped at the same point all the way around the model. Therefore, they make the nose smooth but also give the body tube a glossy, mirror-like surface to reduce friction drag. Experiments conducted by Mark Mercer, then a young rocketeer from Bethesda, Maryland, showed a 24 percent increase in friction drag between a rough nose and a smooth nose. To overcome the effects of friction drag, it's important that the entire surface of a model rocket be as smooth and glossy as possible.

The impact of tiny air molecules on the surface of an object, such as your hand sticking out of the window of the moving car, creates a drag force known as *pressure drag*. Obviously, this pressure on the front end of an object moving with respect to the air is caused by the impact of the air molecules on the object. But it's also possible to have *negative* pressure, a region of fewer air molecule impacts than the surrounding air. (Remember that the body is totally immersed in the earth's atmosphere, which produces a nearly constant, static, ambient environmental pressure of about 14.7 pounds per square inch at sea level.) The region *behind* your hand in the auto airstream has *negative* pressure or *partial vacuum.*

If no boundary layer existed and if the air had no viscous characteristics to make it act like a very thin syrup, pressure drag would not exist. This condition is shown in Figure 10-4. The air would move smoothly apart in front of the object, allow the body to pass through it, then close in quietly and completely behind the object with hardly a ripple. This is the case only for very small objects such as raindrops moving at very low speeds through the air. Or for small flying

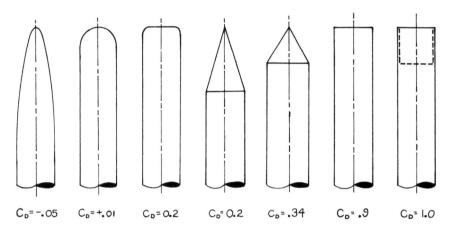

$C_D = -.05$ $C_D = +.01$ $C_D = 0.2$ $C_D = 0.2$ $C_D = .34$ $C_D = .9$ $C_D = 1.0$

FIGURE 10-5   Pressure drag coefficients for various nose shapes.

insects. (Not long ago, ignorance of the boundary layer and air viscosity led to the erroneous contention that, theoretically, bumblebees couldn't fly. The fact that bumblebees do indeed fly prompted scientists to search for the reasons.)

An object the size of a model rocket moving through the air at 250 feet per second simply doesn't give the air enough *time* to slip out of the way and close in again behind it. The air must be *shoved* aside by the nose, an action that creates pressure drag and a boundary layer. If you've done a good job of design and construction, the boundary layer becomes turbulent at the base of the nose and finally breaks away completely from the surface of the model at the blunt rear end, where the air goes into swirling, eddying motion, completely unat-

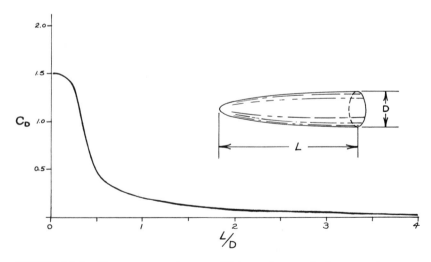

FIGURE 10-6   The pressure drag coefficient of a parabolic nose shape as a function of its length-to-diameter ratio.

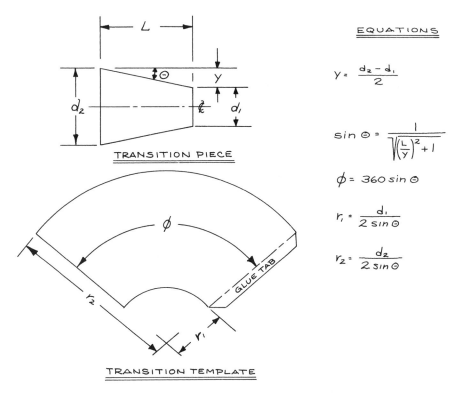

EQUATIONS

$$y = \frac{d_2 - d_1}{2}$$

$$\sin \Theta = \frac{1}{\sqrt{\left(\frac{L}{y}\right)^2 + 1}}$$

$$\phi = 360 \sin \Theta$$

$$r_1 = \frac{d_1}{2 \sin \Theta}$$

$$r_2 = \frac{d_2}{2 \sin \Theta}$$

TRANSITION PIECE

TRANSITION TEMPLATE

FIGURE 10-7   How to lay out a conical boat tail or transition piece having any fore and aft diameter, angle, and length starting with a flat sheet of material.

tached to any object. Thus, a *wake* is created, just like that behind a boat moving in the water. The wake is the low-pressure area that tries to retard the motion of the model, and it must therefore be considered as a form of pressure drag. It's called the *base drag* and is part of the total pressure drag of the model. For many model rocket designs, base drag is the major portion of the total pressure drag on the model.

To reduce base drag, many model rocket designers make a *boat tail* from stiff card stock paper, laying it out in accordance with the principles illustrated in Figure 10-7. This shows you how to make a tapered boat tail from a flat sheet of paper. You can also use these instructions to make a *transition piece* called a *shoulder* where the body diameter of the design increases. The equations allow you to lay out a transition piece or boat tail with angles and lengths of your choice.

Additional drag studies done by George Pantalos in the Ohio State University wind tunnel in 1973 showed that there's an optimum place to put a boat tail or transition. Pantalos tested three basic model rocket shapes shown in Figure 10-8. Model A is a typical large-diameter model rocket with a constant body diameter. Model B is Model A with a boat tail at the rear of the

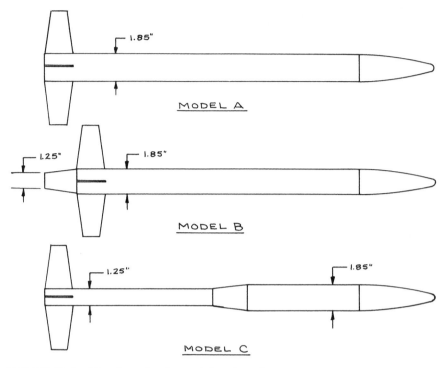

FIGURE 10-8   The three basic model rocket shapes tested for overall drag by George Pantalos at Ohio State University.

model to reduce the diameter from that of the body to that of the motor. Model C has its diameter reduction or transition right behind that section of the model requiring the large-diameter tube for payload-carrying purposes.

Which model had the least amount of drag when tested in the wind tunnel? If you chose Model C, you're correct. It has the *least body surface area* and therefore the least amount of friction drag. Model B has the same base area and base drag as Model C but greater surface area and therefore greater friction drag. Model A, of course, has more base drag than Model B and Model C, and more surface area than Model C. See how these different forms of drag interact with one another and how designers must make technical trade-offs?

Interference drag is caused by the interruption of the boundary layer airflow over the body and the fins by the junctions between the body and the fins—that is, it's caused by the interference of the flow between the two surfaces. Technically, this might be considered part of pressure drag, but we separate it from pressure drag in model rocketry because we face the question: Which model will have the least amount of interference drag: one with three fins in triform configuration or one with four fins in cruciform configuration? The obvious answer is that the model with three fins will have 25 percent less interference drag than the model with four fins. This is why most high-performance model rockets have three fins.

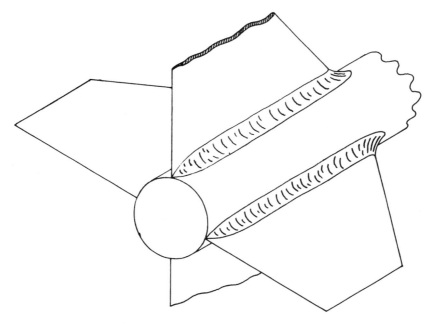

FIGURE 10-9   Fillets of glue or putty at the intersection of the fin roots and the body tube help to reduce interference drag. Fillets are exaggerated for clarity.

This leads to an interesting story. The first three-finned rocket was the historic WAC-Corporal designed by the Jet Propulsion Laboratory of the California Institute of Technology in 1945. Before that, all rockets and bombs had four fins. Some aerodynamic experts didn't believe that the three-finned WAC-Corporal rocket would be stable in flight. Dr. Frank Malina, the rocket's designer, quietly pointed out that arrows with three fletching feathers had been flying in a stable fashion for centuries. Often, it pays to bridge the apparent gulf between two seemingly unrelated fields of human endeavor.

Interference drag can be reduced by the use of *fillets* as shown in Figure 10-9. The optimum fillet radius should be 4 percent to 8 percent of the fin root chord. A fillet that's too large will increase the surface area and hence the friction drag. A satisfactory fin fillet for most small model rockets can be made with a bead of glue.

Another method of reducing interference drag was suggested by Dr. Gerald Gregorek of Ohio State University and confirmed by wind tunnel tests: move the fins forward of the aft end of the body tube by a distance of about one body diameter. (But be sure to check the CG-CP relationship when you do this.) This slight forward location of the fins permits the airflow to smooth out along the body tube behind the fins before it encounters the turbulent region of the model's base.

Parasite drag is drag caused by anything that sticks out from the body to interrupt the smooth flow of the boundary layer over the model. A major source of parasite drag of a model rocket (which is an exceptionally clean aerodynamic shape, by the way) is the launch lug that's glued to the body tube about halfway along the body. On an otherwise clean, streamlined, well-made model rocket, the launch lug's parasite drag can amount to as much as 35 percent of the total drag on the entire model. Obviously, the way to reduce this sort of parasite drag is to eliminate the launch lug by using a tower or piston launcher. However, a suitable compromise is to locate the launch lug in the fin-body joint of one of the fins; this reduces the launch lug's parasite drag to about 20 percent of the total drag of the model. The effort of relocating the launch lug to reduce drag by 15 percent is certainly worthwhile.

Induced drag is drag created by lifting force, or drag *induced* by the lifting characteristics of a surface such as a model rocket's fins. A body tube and nose combination doesn't generate enough lift to really matter. In fact, Barrowman correctly ignored nose-body lift in his CP calculation method. This is because aerodynamic studies of the lift characteristics of bracing wires and struts on World War I biplanes indicated that such lift was negligible. The major lift-producing parts of a model rocket are its fins. They don't produce lift as you might normally consider it—lift *against* the force of gravity. They produce lift in the true and classic definition of the term: an aerodynamic force *perpendicular* to the surface, regardless of the orientation of that surface with respect to gravity. In model rocketry, this lift force is the aerodynamic stabilizing or restoring force we discussed in the last chapter.

Drag due to lift—induced drag—is produced by any surface that generates lift. The reason for this is best understood by looking at Figure 9-4 where you can see that any surface generates drag as well as lift. But some surfaces can be designed to produce less induced drag because of two factors: (1) their cross-sectional or airfoil shape, and (2) their planform shape.

Without going into the mathematical details, lift is generated by a surface at an angle of attack because high pressure exists on the side most directly exposed to the oncoming airflow and low pressure exists on the other side. Lift is therefore caused by a *difference in pressure* on the sides of a shape. Even a flat plate generates lift when at an angle of attack. But specially designed shapes called *airfoils* generate more lift at lower angles of attack. They are able to continue generating lift at high angles of attack. And they possess a higher lift-to-drag ratio. (Model rocket fin airfoils are symmetrical airfoils and do not produce any lift at zero angle of attack.)

The most important source of induced drag is the airflow around the *tip of the fin*. If a fin is at an angle of attack, thereby generating lift to stabilize the model and restore it to zero angle of attack, high pressure exists on one side of the fin and low pressure on the other side. The high pressure spills over the fin tip in an attempt to relieve the low pressure on the other side. This creates a *span-wise flow* over the entire fin, more noticeably near the fin tip. Because the air is also flowing backward over the fin surface, the result is the creation of a corkscrew or helical motion of the air known as a *vortex* that is shed from the fin tip (see Figure 10-10). Energy is required to create and maintain this vortex, and this energy loss shows up as induced drag. The smaller and weaker the tip vortex, the less the induced drag.

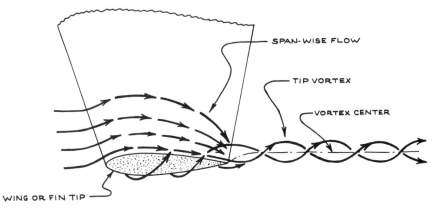

FIGURE 10-10   The creation of the tip vortex of a wing or fin that is generating lift. This is the source of induced drag.

I hear a loud young voice saying, "Well, just put another surface over the tip of the fin, making a tip plate to keep the air from spilling over the tip in the first place." This is certainly one solution to the problem. However, it requires very careful design trade-offs so that the reduction of induced drag isn't greater than the additional friction drag of the plate and the interference drag at the joint of the fin tip and the tip plate. The use of tip plates has paid off handsomely on some large jet transports where specially designed tip plates called Whitcomb winglets are used. You'll see them on the McDonnell Douglas MD-11 and Airbus A320 airliners, for example.

However, the simplest way to reduce the tip vortex and the induced drag is to shape the fin tips properly. The induced drag of several styles of fin tips is shown in Figure 10-11. Again, the data come from wind tunnel tests made to determine the best shape for airplane wings. After all, a model rocket fin can also be considered as nothing more than a small wing. You might not suspect that an absolutely square fin tip would have the lowest induced drag of all the shapes tested, but it does. Several airplanes use this tip shape, most notably the fast, streamlined, single-engined Mooney M-20 series. However, a sharp tip also has low induced drag. This is also evident from looking at the wing tips on many small, general aviation airplanes. Both tip shapes hinder the spillover of high pressure to relieve the low pressure on the other side, thus reducing the size and strength of the tip vortex. It's also possible to use a conical tip that bends down to create a semitip plate; this works well on the asymmetrical airfoils of aircraft and rocket gliders but not on the symmetrical airfoils of ordinary model rockets.

The *planform shape* of a fin also has a great deal to do with its induced drag. Several common model rocket fin shapes or planforms are shown in Figure 10-12.

The swept-back fins that look so good on a model rocket are actually the *worst* when it comes to induced drag. Swept fins can also exhibit *divergent*

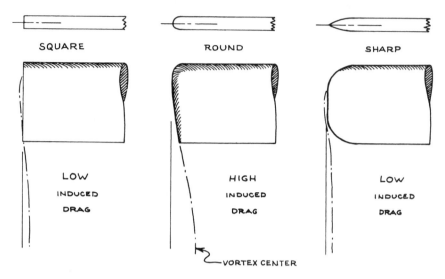

FIGURE 10-11   Induced drag of three simple fin-tip shapes.

flutter. In the case of large swept fins on large model rockets or the swept wings of jet aircraft, the flexure of swept wings at high angles of attack can cause them to begin to flutter. This flutter gets worse within a fraction of a second to the point where the fin breaks or the wing comes off. Swept fins are used only for appearance sake or as a last resort to move the CP to a more rearward position.

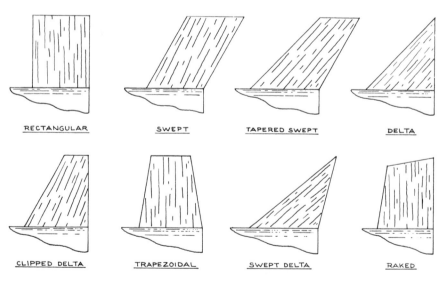

FIGURE 10-12   Some common fin shapes (planforms) used in model rocketry.

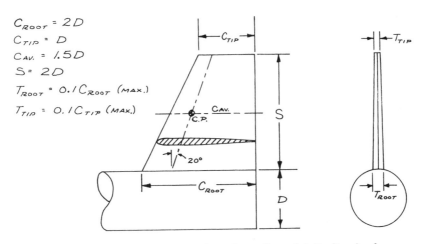

FIGURE 10-13 How to lay out a low-drag clipped delta fin planform.

The two best low-drag fin planforms for model rockets are the *clipped delta* and the *elliptical*. The clipped delta is shown in Figures 10-12 and 10-13. The elliptical planform looks like that of the World War II British fighter the Supermarine Spitfire, which had the most aerodynamically perfect wing ever put on an airplane. However, the clipped delta is easier to make and was used on such equally high-performance airplanes as the Douglas DC-3 and, with no leading-edge sweep, the North American P-51 Mustang.

The difference in induced drag between the clipped delta and the elliptical planforms is about 1 percent in favor of the elliptical. But the clipped delta is easier to make. Therefore, I'm personally partial to the clipped delta and have designed many high-performance, record-setting model rockets using this planform.

The basic design parameters for the optimized clipped delta fin planform are shown in Figure 10-13. These are based on the diameter of the body tube as a point of departure. A clipped delta fin designed using these rules will be slightly oversized for most model rockets. This gives you a margin of safety if you don't care to run the CP calculation. If you do and discover you can use a smaller fin, reduce the C_{root} and C_{tip} dimensions. If you need *more* fin area, increase the S dimension, the fin span. Increasing the fin span doesn't move the fin CP aft, but it greatly increases the fin moment, which will move the model's CP rearward while maintaining low-planform-induced drag.

Now we need to put all of this aerodynamic information together into the design of a low-drag, high-performance model rocket. A typical idealized low-drag model rocket design is shown in Figure 10-14. It has a parabolic/elliptical nose with a length-to-diameter ratio of 3. It has three fins with clipped delta planform. The launch lug is nestled into one of the well-filleted fin-body joints. The aft end of the body tube could not use a boat tail because the motor fits right into the body tube. So the aft end of the body tube is slightly rounded with fine sandpaper and a little bit of the motor casing

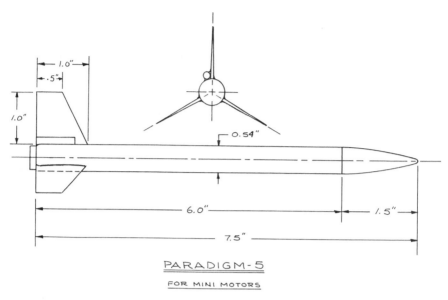

PARADIGM-5

FOR MINI MOTORS

FIGURE 10-14   A typical low-drag, high-performance model rocket design for use with minimotors.

extends from it. The extension of the motor casing beyond the aft end of the body tube serves two functions: (1) It makes it easy to grasp the motor casing to extract it from the model after the flight, and (2) the extension acts like an extension of the body tube to reduce interference drag by smoothing the flow aft of the fins. The model should have a shiny, mirrorlike surface with all balsa grain and body tube spirals filled. To prove that this idealized model rocket is something that can be built, Figure 10-15 shows some competition and sporting designs.

The drag coefficient (C_d) of an idealized model rocket such as that in Figure 10-14 has been calculated at 0.4 at a velocity of 200 feet per second. Flight tests with models that have been tracked for altitude have indicated that the C_d can be as low as 0.25.

The drag coefficients of various parts of an average, well-designed model rocket have been calculated by Gregorek and Pantalos as follows:

Nose and body pressure and friction drag = 0.2 (50 percent of total)

Base drag = 0.06 (15 percent)

Fin friction, induced, and pressure drag = 0.07 (17.5 percent)

Interference drag = 0.02 (5 percent)

Launch lug parasite drag = 0.05 (12.5 percent)

These numbers have been confirmed by building models, flying them, and tracking them on six to ten flights. The actual altitude data are then compared

FIGURE 10-15 Some low-drag, high-performance contest model rockets. Fins are made from a very thin, semitransparent fiberglass sheet.

against altitudes calculated with RASP-93 for various values of C_d. The closest agreement between actual tracked altitude and computed altitude then provides a very good estimate of C_d.

The reason there are so many model rocket designs is because model rocket designers make the various trade-offs differently. Any model rocket designer must make technical compromises. The way a particular model rocket designer does this results in a model rocket whose appearance is as distinctive as a signature because of the particular trade-offs and compromises favored. You'll probably develop your own design signature over the years.

A lot remains to be learned about model rocket aerodynamics. One of the hottest areas of controversy centers on aerodynamics, just as in model aeronautics and even in full-size aircraft and spaceships. All of the data haven't come in yet. Much remains to be discovered. New materials constantly allow us to try new shapes and concepts. A lot of guessing goes on because many hypotheses aren't supported by hard data. And much of this hasn't been reduced to meaningful and useful design rules. All of this indicates that model rocket aerodynamics remains an area ripe for original research that doesn't require an operating knowledge of advanced mathematics, expensive equipment, precision measuring devices, and years of education. Model rocketry aerodynamics is full of fun and games. Just when you think you've got the whole answer, some other model rocketeer comes along with a design developed from other data—and the controversy is under way.

Don't think that mannered, reasoned controversy is a bad thing. At a 1957 space conference, the pioneer fluid dynamics expert and rocket scientist Dr. Theodore von Kármán was asked to sum up the meeting. He rose to his feet and told us, "A very fine meeting. Very well organized. Excellent papers on pertinent subjects. Good presentations with solid data. *But no arguments!* Ladies and gentlemen, how can we possibly have progress without controversy?"

So, if you have the data to support your opinion, argue away. In the process, you'll have fun and learn a lot more. Someday, you may graduate to designing big things that fly with people in them. Including spaceships.

11

MULTISTAGE MODEL ROCKETS

Thus far, we've discussed only single-stage model rockets. There are two reasons for this. First, nearly everything that applies to single-stage model rockets also applies to multistage model rockets—and sometimes even more so. Second, unless you understand the principles of single-stage model rockets that we've discussed already, you're going to have considerable difficulty with multistage model rockets.

Multistaging permits you to make a significant improvement in the performance of a model rocket. Multistaging increases the model's total impulse and decreases its final burnout weight by discarding as much unneeded weight as possible. Technically, a multistage model is one in which two or more motors operate sequentially in flight, then the expended motor(s) and the associated airframe(s) are discarded or jettisoned after burnout. See Figure 11-1.

With a single-stage model, you can increase the burnout velocity by decreasing the weight of the model until the model is at the optimum weight—not so light that it starts behaving like a feather, and not so heavy that the motor thrust can't accelerate it enough. Or you can use supercareful streamlining and drag-reduction techniques. Or you can increase the total impulse of the motor. However, these methods have limits beyond which you cannot go because you run up against technical, constructional or operational limitations.

Multistaging offers a method of increasing burnout velocity without some of the disadvantages and limitations of single-stage models—but with some nasty little tricks of its own that you must watch out for! The simplest form of staging is *series staging,* shown in Figure 11-2.

Essentially, the payload of the lower stage is itself a model rocket: the upper stage. A motor and its enclosing partial airframe—body tube, fins, and motor mount—make up the lower or first stage, often called a *booster.* The upper stage is a complete airframe, a single-stage model rocket. Ignition of the lower-stage booster motor accelerates the entire multistage model into the air. At lower-stage motor burnout, the booster assembly separates from the upper

FIGURE 11-1   Multistaging offers a way to increase performance but adds complexity to model rocket design and flying.

stage and the motor of the upper stage is ignited. This adds the total impulse of the lower-stage motor to that of the upper-stage motor, which in turn adds the velocity imparted by the lower stage to the velocity that can be attained by the upper stage. Since we learned that the peak altitude is a function of the square of the burnout velocity and since the burnout velocities of the two stages are added together, you can quickly see that the peak altitude of a multistage model can be very high indeed. However, aerodynamic forces can also be very high.

In addition to increasing the total impulse of the model, series staging also decreases the final burnout weight because the expended lower-stage motor casing and airframe are jettisoned when their usefulness is at an end. Separation of a series-stage model occurs with no apparent visible delay. Actually,

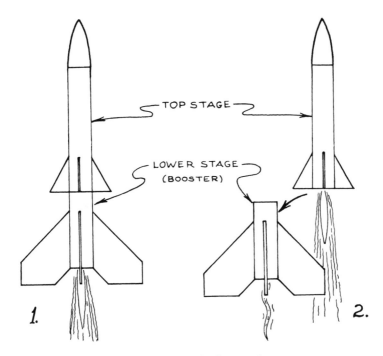

TOP STAGE

LOWER STAGE
(BOOSTER)

1.

2.

FIGURE 11-2   Series staging.

there is some, as we'll find out. Separation occurs when the lower stage has imparted its maximum velocity to the upper stage.

Stage separation shouldn't occur when the model reaches the *maximum altitude* to which the lower stage will carry it. Staging should happen at *maximum velocity*. Why is series staging done at maximum velocity rather than at maximum staging altitude? An example will show why.

Let's use a hypothetical model rocket whose lower stage, without separation and ignition of the upper stage, can carry the entire model to a burnout velocity of 100 feet per second and a peak altitude of 250 feet, as shown in Figure 11-3. Suppose that the upper stage all by itself can achieve a burnout velocity of 100 feet per second and a peak altitude of 250 feet. If staging took place at the *peak altitude* of the lower stage, the peak altitudes would add together, giving a total peak altitude of 500 feet for the combination staged in this fashion.

However, if the upper stage is separated and ignited at the *maximum velocity* of the lower stage, the 100-foot-per-second velocity of the lower stage is added to the 100-foot-per-second of the upper stage, giving a final burnout velocity of 200 feet per second for the staged combination. Peak altitude is therefore four times what it would be at a burnout velocity of 100 feet per second, and the upper stage reaches a theoretical apogee of 1,000 feet. Therefore, a model rocket staged at maximum velocity of the booster will go roughly

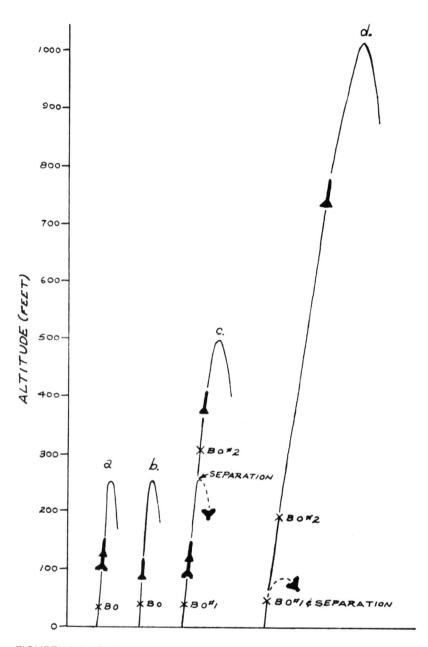

FIGURE 11-3 Series staging at various points in a model rocket flight. (a) Lower stage only ignites. (b) Top stage only. (c) Model stages at peak altitude of lower-stage flight. (d) Model stages at lower-stage burnout (BO) point and therefore at maximum-stage velocity.

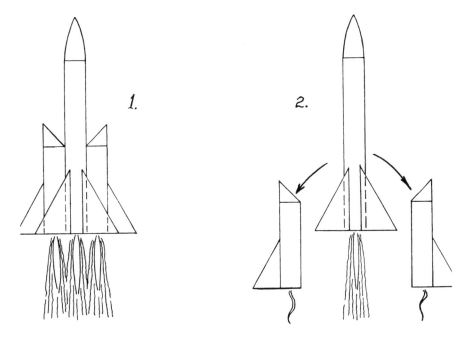

FIGURE 11-4   Parallel staging.

twice as high as one staged at the maximum altitude of the booster—neglecting aerodynamic drag, of course.

Another type of staging is called *parallel staging,* because the stages operate in parallel rather than in series. As shown in Figure 11-4, the parallel-stage model rocket leaves the launch pad with all motors of all stages operating. The stage motors and airframes are quite properly called boosters and the upper stage is the core or the sustainer.

The NASA space shuttle and the Atlas ICBM are parallel-stage rocket vehicles. For parallel-stage model rockets, the booster motors are selected to have less duration than the sustainer or core motor. Perhaps the sustainer motor is a Type C6 while the booster motors are Type A6s. At booster motor burnout, the booster airframes are separated by air drag or other techniques, leaving the sustainer motor still thrusting and accelerating the model's core vehicle.

Parallel staging offers very high thrust and acceleration at liftoff. It also eliminates the problem of air start of series-stage upper-stage motors. However, ignition of multiple or clustered motors is also fraught with the possibilities of not igniting all motors on the pad. Modern low-current igniters have helped solve a lot of headaches in clustered ignition. For more details on clustered motor ignition, see chapter 14. And refer to the various technical reports published by the manufacturers that deal with cluster ignition of their particular motors.

For many years, it was believed that parallel staging wasn't practical for model rockets because of all the difficulties involved. Pat Artis of Ironton, Ohio, perfected parallel-stage model rockets and demonstrated them at NARAM-7 in 1965. By 1966, Artis had refined his designs to carry the booster pads up near the nose of the core model, thereby bringing the CG of the model well forward during the critical boost phase and eliminating the need for large fins to ensure stability during the boost phase. Artis showed that parallel-stage models had superior performance for payload carrying because of their high liftoff acceleration. This permits a heavy payload model to get airborne with an airspeed that ensures stability.

However, since then, there hasn't been much activity in parallel staging. Perhaps this is because a lot of model rocketeers don't know about parallel staging. Although there are some parallel-stage model rocket kits available, the series-stage model rocket is the most commonly flown type.

Some model rocketeers love air starts. This isn't multistaging. I don't recommend it. The air-start technique ignites one or a cluster of motors on the pad. Ignition of one or more other motors in the model is delayed by various complex ignition schemes. The model takes off with one set of motors operating, then another set ignites in the air—you hope. Because the initial motors aren't dropped or jettisoned, the model has to carry their empty weight throughout its flight, lowering the model's performance. The only benefit gained by an air start is to increase the overall burning time of the model's propulsion system; this advantage is heavily outweighed by the additional drag, additional weight, increased complexity of the delayed ignition methods, and reduced reliability of air starts. Some modelers refer to air-start models as "ego boosters."

Success in flying a series-stage model rocket depends on careful design, construction, and operation. This, in turn, depends on understanding what happens during a series-stage model's flight. This is because air-start ignition of the upper-stage motor can pose a problem if it isn't done carefully, especially in a model of your own design.

Ignition of the first-stage motor of a series-stage model rocket is accomplished electrically in a manner identical to that of a single-stage model rocket. A series-stage model rocket also uses a launch pad. If the stage model is longer than 24 inches or weighs more than 6 ounces, a $3/16$-inch-diameter launch rod 5 to 6 feet long should be used because liftoff acceleration will be low and the airspeed leaving the launcher may be low as well. Figure 11-6 shows a cross-section of a typical two-stage series-stage model rocket at various times during its flight.

Drawing A shows the model shortly after liftoff; the propellant in the lower-stage motor is burning forward. Note that the lower-stage airframe is nothing more than a hollow body tube with motor mounts installed and fins attached. The body tubes of the two stages are simply butted together. The body tubes are often kept in alignment either by the upper-stage motor extending slightly into the lower-stage booster as shown, or, with models of diameters larger than the motors, by a paper collar extending from the booster

FIGURE 11-5   Pat Artis (left) describes his parallel-stage model rocket to Cal Weiss of NASA. Note forward location of booster pads.

stage up inside the upper stage to provide support and alignment. Thrust of the lower stage is transferred to the upper stage through the lower-stage motor mount, the lower-stage body tube, and the butt joint between the two stages. The lower-stage or booster motor should be firmly held in the booster body tube with a motor clip or with a paper ring *inside* the aft end of the booster body. (You have to insert the lower-stage motor from the front of the booster stage if you use the aft-ring motor retention method, so this is good only for short booster stages.)

If the model has a body diameter larger than that of the motors, an internal stuffer tube is used. A stuffer tube is essentially an extension of the lower-stage motor mount tube forward to the upper-stage motor. The stuffer tube prevents the expansion of the blow-through gases of the lower-stage motor, as we'll see in a moment.

In drawing B, the lower-stage motor has almost reached burnout. Only a thin disk of propellant is left in the lower-stage motor. Separation and staging are only a split second away.

In drawing C, the hot combustion gases under pressure have broken through the thin disk of propellant and are sending hot chunks of burning lower-stage motor propellant forward. These pass up the nozzle of the upper-stage motor, causing it to ignite.

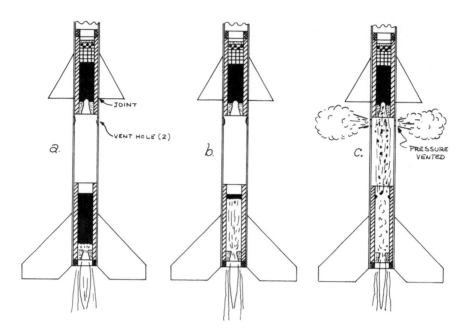

FIGURE 11-6   Cross-section diagram of a multistage model rocket at various progressive points of lower-stage operation. The lower-stage vent holes relieve the internal pressure of the lower-stage motor blow-through and prevent the lower stage from separating before the upper stage ignites.

Once the upper stage has ignited, its jet pressurizes the lower-stage body and blows it off the model. The motor clip or the aft ring prevents the lower-stage motor from being ejected from the booster airframe during this process.

If the lower-stage motor casing is ejected and the booster airframe *isn't*, the jet from the upper stage undergoes an unusual phenomenon known as *overexpansion* or the Krushnic Effect. It blasts down through the lower-stage body tube, ruining the booster airframe. This overexpansion of the exhaust gases reduces the upper-stage thrust to a couple of ounces, meaning that your model won't go very high but will probably deploy its recovery system before it lands.

If staging is successful—and it can be 99.999999999 percent successful if everything's done right—the upper stage will then keep going up on its own, leaving the lower-stage airframe to tumble back to the ground where it lands unharmed. Lower stages are deliberately designed to be unstable all by themselves so that they do tumble. This takes some tricky figuring with the Barrowman Method.

The biggest problem with multistaging is ensuring the ignition and air start of the upper stage of a series-stage model. The problem manifests itself as follows: the model comes up on stage separation and the blow-through feature separates the stages before the hot pieces of booster motor propellant have time to move forward and go up the nozzle of the upper-stage motor to ignite it. When this happens, the upper stage doesn't ignite and therefore cannot eject the recovery device. Although the booster tumbles back safely, the upper stage performs what is known as a death dive.

It took a long time to find out exactly what goes on at stage separation and how to ensure upper-stage ignition. Various techniques were developed to delay the separation of the booster airframe until the upper stage had positive ignition and could blow the booster off. Vernon D. Estes came up with the one that was most common for a long time. The model was designed so that the booster is short and the two motors butt together. The motor casings are secured together with a single wrap of cellophane tape. This keeps the motors together during staging until the upper-stage ignition blows the motors apart at the taped joint. The drawback to this positive method of air-start ignition is the requirement that the motors butt together. This results in short, underdamped model designs that require very large booster and upper-stage fin areas, often at the expense of drag, weight, and reduced dynamic stability.

The alternative was to tape the stage airframes together instead of the motors. This permits lengthening the booster-stage airframe, resulting in smaller booster fins and better dynamic stability. However, sometimes the tape stays with the upper stage, creating excessive or asymmetrical drag.

A thorough investigation of what happens at staging was carried out by Arthur H. "Trip" Barber III while he was still an undergraduate at MIT. He ran series-stage motors taped together in a static test stand. The data showed that the upper-stage motor didn't ignite immediately after blow-through and burnout of the lower-stage motor.

There was a delay of 0.03 second—$^{30}/_{1,000}$ of a second or 30 milliseconds— between burnout/blow-through of the lower-stage motor and the ignition of the upper-stage motor. This was repeatable, test after test. Thrust dropped to zero for 30 milliseconds at staging.

I came upon the solution to the series-staging air-start ignition problem as a result of a series of circumstances that are typical of how scientific and technical problems get solved. I'll go through it here to indicate to you how some of the everyday problems of science and technology are approached and conquered.

For almost a decade, I'd been thinking about this staging problem and making some flight tests and static tests. However, I didn't have the high-resolution static test stands that were at Trip Barber's disposal at MIT. When I heard him present his data at a model rocket convention in San Jose, California, in 1977, I knew I was on the right track. Trip had provided the one missing piece of datum that I needed to proceed.

I'd been thinking conceptually about what really happens at staging. I'd hypothesized as follows: at the instant of breakthrough of the lower-stage motor propellant disk, the lower-stage body tube suddenly becomes full of two things: (1) hot combustion gas and (2) burning pieces of fractured propellant from the lower-stage motor. The hot combustion gas is less dense and has less total heat content than the burning pieces of propellant and it moves forward through the booster body tube at the speed of sound, creating a rapid overpressure front known as a shock wave. The hot pieces of burning propellant are heavier and therefore move more slowly. The hot gas doesn't have enough heat energy to ignite the upper stage motor; only the hot pieces of propellant can do that. But by the time the propellant pieces get to the upper end of the body tube, the shock wave of hot combustion gas probably has already blown the booster stage off the model.

Everybody who flew series-stage models knew that you had to keep the stages together until the ignition of the upper-stage motor blew them apart. This was accomplished by taping motors and stage airframes together. But nobody knew why.

Trip Barber's discovery of the delay between booster motor burnout/breakthrough and upper-stage ignition meant that my hypothesis had some merit. It explained *why* motors or stages had to be taped together to delay their separation.

If the shock wave of hot combustion gas was blowing the stages apart prematurely, I reasoned that we had to get rid of the pressure of the hot combustion gases in the lower-stage body tube. So I vented the lower-stage body tube by punching *two* 1/4-inch-diameter holes on *opposite sides* of the tube just below where the nozzle of the upper-stage motor would be. I used two vent holes because I didn't want the venting combustion gas to come out on only one side of the model, thereby creating a side force like firing an attitude thruster. Two vents on opposite sides of the body tube equalized the thrust forces from the venting gas.

The coupling between the two stages had to be tight enough to keep drag forces or any residual unvented gas pressure from blowing the booster stage off. If the coupling was too tight, the jet exhaust from the upper-stage motor would vent through the holes until it literally burned the lower-stage body tube away. So the coupling was made just tight enough that I could pick up the model by the aft end of the booster without having the stages separate.

I built a "standard" upper stage (several of them, just in case) that was equivalent to today's Quest Tracer and a series of boosters of varying lengths. Later, I flew two-stage models with smaller and larger diameters to prove the method.

Over a period of several months, I flew over 100 two-stage flights with 100 percent success in upper-stage ignition. Motors were separated by as much as 12 inches. At first, a lower-stage booster airframe was good for only about six flights because the blast of the upper-stage motor jet that blew the booster off the model also charred the inside of the booster body tube. So I cut a bunch of liner papers from ordinary 20-pound bond paper with dimensions such that I could roll one piece up and insert it into the booster body. It provided an inner layer all the way around and extending from the booster motor mount right up to the lower edge of the vent holes. I used a new liner paper on each flight and the booster airframes then lasted indefinitely.

Other model rocketeers have used the vent method, and it works for them as well. We can now achieve 100 percent air-start ignition of upper-stage motors with boosters up to 12 inches long. We've had the solution to staging now for nearly fifteen years, but you may not see this vent method incorporated into kits or even other modelers' designs. I didn't patent it, so that means the designers and modelers haven't read this book. Old ideas and concepts die hard. When you're flying your first kit-built multistage models, however, follow the instructions of the manufacturer relating to the staging of the kits.

You've probably noticed that the lower-stage model rocket motors for both series staging and parallel staging are special. They have no time delay or ejection charges. Their NAR type numbers end in a dash-zero—B6-0, for example. This allows for staging at burnout of the lower-stage motor.

Do not—I repeat, *do not*—use a standard motor with a time delay and ejection charge in a lower stage. The model will lift off normally, go through lower-stage powered flight, then start to coast. Because a stage model is heavier than a single-stage model, due to its additional stage airframe and motor, the model will coast upward and arc over. Because of the time delay, the lower-stage motor will probably separate the stages with the model pointed *downward*. If the upper-stage motor happens to ignite in spite of the head cap in the lower-stage motor, the upper stage will be under thrust pointed *downward* with the motor thrust assisted by gravity. It will come down *very* fast. There's no time for the recovery device to eject to slow the model nor would it stay in one piece if it did deploy. The upper stage usually impacts under thrust. Very little usually remains of such reentry models. And they are *exceedingly* dangerous.

Safety Rule: *Always* check carefully to make sure you've installed a dash-zero motor in all lower stages.

Although each lower-stage motor must be a dash-zero type, the motor in the upper stage must have a time delay and ejection charge to deploy the recovery device. Because of the higher final burnout velocity of stage models, the upper stage has to coast for a longer period of time before all its momentum is converted to altitude. Therefore, upper-stage motors must have a longer time delay to prevent deployment of the recovery device before peak altitude is reached. Long-delay motors are made and specified for upper stages. When in doubt, however, use a shorter time delay on your first flights to prevent cliffhangers where the model goes over peak altitude and deploys the recovery device on the way down.

As stated earlier, multistage models weigh more at liftoff than single-stage models because of the additional weight of their lower-stage airframes and motors. Therefore, with normal motors, multistage models lift off and accelerate more slowly. Because of these lower launch velocities, good stability is important if a safe and predictable flight is to be achieved by a multistage model. I've seen just about everything happen to a model rocket in flight, but the most frightening is a two-stage model that lifts off, becomes unstable, thrashes around in the air 25 feet above the launch pad, stages, *then becomes stable*, usually pointed *down*! So check stability, *please*!

Multistage models follow the same rules for stability as single-stage models. All of the stability requirements must be met. For a two-stage model, the CP and

CG of the two stages together (in launch configuration with loaded motors in both stages) must be in a stable relationship. The CP and CG of the upper stage alone must be properly located. Thus, stability checks must be made for every flight combination of a multistage model—that is, in its various stage configurations and in its final top-stage configuration. For a three-stage model, three stability checks must be made—for all three stages together, for the top two stages together, and for the top stage all by itself. See Figure 11-7. Although you may occasionally get away with flying a new single-stage design without a stability check, you should *always* make stability checks for multistage models. At the very least, subject the completed model to the simple swing test.

Because of their lower liftoff velocities and larger fin areas, multistage models are *very* susceptible to weathercocking. Therefore, they should be launched in winds of less than 5 miles per hour. If in doubt about the wind, *don't launch* that multistager; wait until the wind dies.

Lower stages are made with the same techniques as those used in building single-stage model rockets. In essence, the upper stage of a two-stager is the nose of the lower stage. Recovery devices are not normally installed in lower stages (although I've done it) because of the need for the blow-through to ignite the next stage. A lower-stage airframe, however, must be built in a more rugged manner than a simple single-stager. It will tumble during descent and land harder, which means that its fins—which are usually larger than those on a single-stage model—have a greater chance of being broken off. Double-glued joints should be used throughout, and fin joints should be reinforced with tissue fillets.

Because lower stages don't deploy brightly colored recovery devices as single-stage or upper-stage models do, they're more difficult to locate after they've landed, even though they probably won't go as far from the launcher. Therefore, lower stages should be painted a very bright fluorescent red or orange so that they can be seen on the ground.

Couplings between stages should be designed and built so that the model will not bend or jackknife in the air at the location of the joint. Most kit models have this sort of solid joint. Numerous different joint designs exist. Basically, the joint and coupling should permit lengthwise separation but prevent wobbling between the stages in the pitch-yaw axes. It's very frightening to have a multistage model jackknife in midair and come apart. If that happens, you're not quite sure when or if the upper stages are going to ignite and where they'll be pointed if they do.

Launch lugs may be placed anywhere on a multistage model. However, here's a tip: put the launch lug in a fin-body joint fillet only on an upper stage. This permits you to fly the top stage as a single-stager if you wish.

To achieve maximum possible stability during boost, don't align the fins of the various stages in a fore-and-aft direction. This puts the lower-stage fins in the wake, downwash, and vortex pattern of the upper-stage fins and greatly reduces their effectiveness. You need all the effectiveness you can get from lower-stage fins. Instead, *interdigitate* the fins. Put them out of line with one another as shown in Figure 11-8.

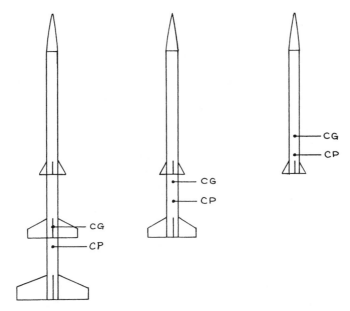

FIGURE 11-7 Stability checks on a three-stage model rocket. Each flight combination must be checked separately as shown.

Upper stages can achieve some very high maximum velocities, sometimes well in excess of half the speed of sound (Mach 0.5). They should be well built with exceptional care devoted to drag reduction if maximum performance is desired. They should also be built strongly because the drag force on an upper stage can exceed the weight of the model by a factor of 10 or more. I've seen upper stages come completely apart after staging, leaving the sky full of fins and other parts. They have reached the legendary speed of balsa.

We've been speaking generally of two-stage models with a few references to three-stagers. The principles apply to three-stage models, too. However, don't try a four-stage or eight-stage model rocket.

Why should you limit your models to three stages? Simply because the reliability of series-stage model rockets decreases according to the inverse of the square of the number of stages. Thus, a two-stage model is roughly one-fourth as reliable as a single-stage model, while a three-stage model is only one-ninth as reliable as a single-stager. This would make a four-stage model one-sixteenth as reliable.

And a three-stage model may actually achieve a *lower* peak altitude than a two-stage model, even when the three-stager has more total impulse. This is because a three-stager is heavier at liftoff and because of air drag may actually have a final burnout velocity that's less than that of a two-stager. If a high burnout velocity is achieved, drag forces may rob the model of most of its altitude during the coasting phase.

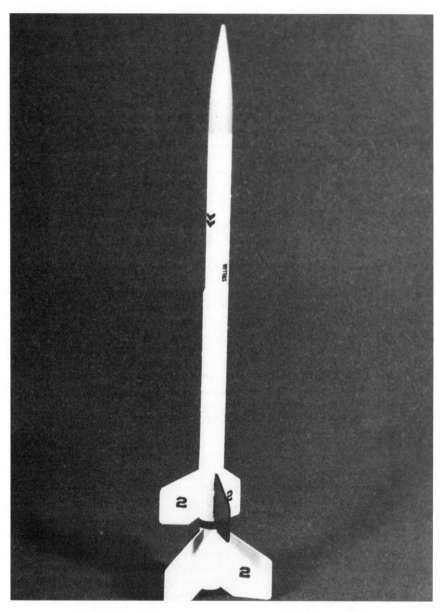

FIGURE 11-8 This two-stage model illustrates interdigitated fins. The upper-stage fins are not aligned with the lower-stage fins, thus making the lower-stage fins more effective.

Performance analyses conducted with a variation of the RASP-93 program have shown that a multistage model rocket with more than three stages will not outperform a three-stager. Usually, it won't outperform a *single-stage* model rocket! Besides, a properly designed and successfully flown three-stage model rocket will probably go completely out of sight. Then what are you going to do? You've probably lost it.

Multistage model rockets are capable of exceptional altitude performance. But, because of the higher potential for something to go wrong, all safety precautions must be rigorously and very carefully followed. Tilt the launch rod a *few* degrees away from the vertical so that the model will land downrange and away from people if it suffers a staging failure.

Multistage models are fun. They're also difficult to build and fly. If anyone tells you that model rocketry is kid's stuff and that you're playing with toys, that person hasn't seen a multistage model rocket climbing skyward at Mach 0.5.

12

RECOVERY DEVICES

Because recovery devices are required on all model rockets, the hobby is virtually free of hazards caused by freely falling objects that are large enough and heavy enough to do damage or cause injury when they return to the ground. The rules of the NAR do not permit the jettisoning and free fall of any part of a model rocket, such as the motor casing, unless the falling part tumbles to slow its speed and uses a streamer or other readily visible surface that can be seen. The model rocket itself must be capable of more than a single flight and must return to the ground so that no hazard is created. If you do everything right, follow all the rules, and read the instructions, you have a 99.99999 percent chance of having your model rocket perform in this safe fashion. If something should go wrong, as it occasionally does, the design of the model and the materials used in its construction are intended to reduce the hazard to acceptable levels.

Although we've discussed recovery devices in general terms earlier, we'll now cover them in detail. Making and using various recovery systems involves all sorts of little hints, kinks, tips, and tricks. A number of highly successful and reliable recovery devices have been developed. The type of recovery device used in a given model rocket depends on many factors: the recovery weight of the model, the kind of payload, the size of the model, the anticipated weather conditions of the flight, etc.

All recovery devices are actuated or deployed by the ejection charge of the solid-propellant model rocket motor. The quick burst of gas from the ejection charge can be made to do a number of things to activate a recovery device.

NOSE-BLOW RECOVERY

I don't recall when the first nose-blow recovery system was used or who used it. It was definitely in use during the summer of 1958 during the flight testing of a large number of simple models near Denver, Colorado. It may have been

FIGURE 12-1   A model rocket always uses a recovery device such as a parachute to lower it safely and gently back to the ground, provided it eludes the clutches of the ever-present rocket-eating trees.

used first by a model rocketeer who took the parachute out of his model to save time during flight preparations or to prevent the model from drifting away in a high wind.

Nose-blow is a very simple recovery method. It derives from a similar technique used from 1946 onward at White Sands Proving Grounds, New Mexico, for aiding the recovery of German V-2 rockets. Normally, a 3-ton V-2 rocket falling from an altitude of 100 miles would dig a very large, deep hole in the desert, destroying all the scientific instruments aboard, as well as the rocket itself. One day, a V-2 happened to come apart on the descending leg of its trajectory and tumbled to the ground. The astounded rocket scientists and engineers recovered most of their equipment intact. Thereafter, the flight safety officer merely activated the ring of explosives around the base of the nose during the descending leg of the flight. The V-2's streamlining and stability were destroyed by this action, and the separated pieces fell at a much lower terminal velocity.

In model rocketry, the same principle lies behind nose-blow recovery, but we do it a bit differently. The nose is tied to the rest of the model by the shock cord. When the nose is separated from the model by the ejection charge, the aerodynamics and stability of the model are greatly altered. The

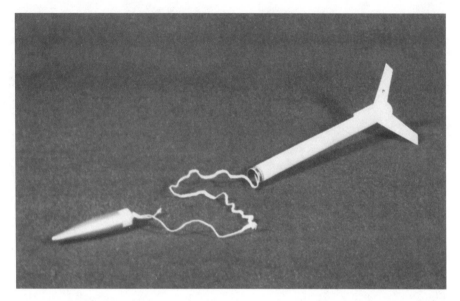

FIGURE 12-2 A nose-blow model rocket uses the motor's ejection charge to separate the nose from the body, thus destroying the stability of the model.

CP-CG relationship is changed, so the model becomes unstable and flutters end over end. With the nose tied to the model by the shock cord, the two parts land together. This eliminates the need to search for both pieces separately. A typical nose-blow model is shown in Figure 12-2.

Although a nose-blow model falls slowly enough to catch with your bare hands, it can still land hard enough to break fins if the landing site is an asphalt or concrete area. Therefore, most nose-blow recovery is used only for models weighing less than 2 ounces (60 grams). It is also used for high-altitude models when you don't want them to drift into the next state. However, you occasionally see nose-blow used on large models with a high area-to-weight ratio. Nose-blow is almost never used when flying a fragile payload.

STREAMER RECOVERY

When model rocketeers wish to slow their model a bit more than is possible with nose-blow recovery and also wish to see it more clearly against the sky and on the ground after landing, they add a streamer to the nose-blow recovery model. This creates a streamer recovery model (see Figure 12-3). Or they may build it that way from the start.

A streamer is a brightly colored strip of plastic film or crepe paper attached to the shock cord or the nose base. The streamer is attached only at one end, although it's sometimes attached in its middle. Streamer dimensions vary between 1 inch and 4 inches in width and 12 inches and 48 inches in length. Many modelers, myself included, prefer to use orange crepe paper

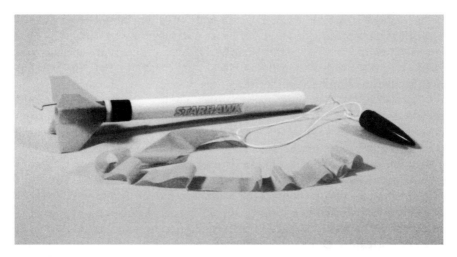

FIGURE 12-3 A streamer recovery model rocket has a paper or plastic streamer attached to the nose base or to the shock cord near the nose.

instead of plastic because crepe paper doesn't get stiff in cold weather, seems easier to fold and roll, does not melt if some ejection charge gas seeps around the recovery wadding, and offers good drag characteristics.

Research done by Trip Barber and others at the MIT Rocket Society indicates there is an optimum streamer size for every model rocket design. Generally, the optimum streamer size to obtain the slowest descent rate has a length-to-width ratio of 10:1. In other words, a 1-inch-wide streamer should

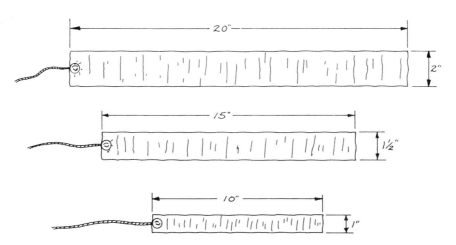

FIGURE 12-4 The best streamer configuration is a length-to-width ratio of 10 to 1 as shown.

be at least 10 inches long and a 2-inch-wide streamer should be at least 20 inches long. According to Barber's tests, nothing is gained by going to a length-to-width ratio of greater than 10 because a longer streamer actually begins to stream rather than flutter. To slow the model down, the streamer must flutter and flap back and forth.

Barber's work also indicates that it is possible to match the streamer dimensions to a given model rocket design to achieve the lowest possible descent rate and therefore the longest possible flight duration time. For this reason, the NAR has placed no limits on streamer size or material and has not attempted to adopt a standard streamer for its Streamer Duration Competition. This category of competition was originally intended as a simplified achieved-altitude contest for beginners and for fun meets. However, like other kinds of competition, it quickly evolved into a highly sophisticated contest category with its own strategy and tactics. In spite of this, however, Streamer Duration remains the simplest of all competition categories to fly.

A streamer must be protected from the hot gas of the model rocket motor's ejection charge. A plastic streamer can be melted and a crepe paper streamer can be burned by the ejection charge gas. Protection is simple and easy: stuff a small wad of loosely packed flameproof tissue down into the body tube and put the rolled-up streamer in on top of it. The wadding blocks the ejection charge gas from the streamer and also acts like a piston to help eject the streamer. The wadding is ejected from the model after the streamer comes out. Recovery wadding must *always* be flameproof material so that there's absolutely no chance it may be ignited by the ejection charge gas and fall smoldering to the ground where it could start a fire in dry grass.

A streamer is packed into a model by first rolling it tightly into a long cylinder. Some competition model rocketeers use an accordion fold. Everyone seems to have a favorite folding method. Often a single line of string or thread is used to attach the streamer to the base of the nose. This string can also be wrapped around the rolled streamer.

The streamer should slide easily into the body tube so that it cannot be jammed when it's ejected later. It should be capable of being ejected easily by the ejection charge. If you wad it in, the ejection charge may not be able to move it.

Don't push the streamer down into the front of the body tube more than about an inch, leaving just enough room to put the shock cord in atop it, and place the nose on the tube. You want all the weight you can get up front to improve the CG-CP relationship.

A streamer recovery device will cause the model rocket to drift with the wind. It will drift farther than a nose-blow recovery model. Often the model lands in a tree. In fact, it's a foregone conclusion among experienced model rocketeers that a model rocket is certain to land in the top branches of any tree within sight of the launch area at some time during the day's flying activities.

Most Streamer Duration competition models have the shock cord attached externally at the CG of the body/fin/motor casing and the nose attached to the shock cord near the end where the streamer is attached. The shock cord lies along the external surface of the model until recovery deployment. This is to make the model descend in a horizontal position, where it offers the increased air drag, thereby decreasing the descent rate and increasing the flight duration time. In these competition models, the shock cord is an 18-inch to 24-inch length of nonelastic carpet thread. A notch in the nose shoulder to clear the shock cord as it emerges from the front of the body tube is cut into the front of the tube. The cord lies back along the external surface of the body tube.

Streamer Duration competition has gotten so sophisticated on the national and international level that many competition modelers use a wide variety of streamer materials such as tracing tissue, Mylar plastic film, or bond paper. Many competitors change the streamer for each flight because they've discovered that a used streamer may deploy with less drag than a fresh one. This only goes to prove that even a competition category that starts as a simple beginner's altitude contest can evolve over a period of only a few years into a very complex event with highly specialized models and unique techniques and methods for flying it. But this doesn't keep Streamer Duration from being an excellent contest for beginners or for fun.

PARACHUTE RECOVERY

Parachute recovery of rockets is an old and established art. It may have been used by the Chinese, the inventors of the parachute itself, in their early fireworks rockets. But the first reported use in the literature is a description of a fireworks demonstration conducted by the Ruggieri brothers in Paris shortly after the French Revolution. (The Ruggieri company is *still* in the fireworks business and is one of the best in the world.) Dr. Robert H. Goddard successfully recovered many of his rockets with parachutes. Rocket flights were made by the American Rocket Society (now the American Institute of Aeronautics and Astronautics) in the 1930s using parachute recovery. Some of these test rockets had replaceable motors like today's model rockets. Many modern research rockets regularly use parachutes for the recovery of the payload or even the entire rocket vehicle. Each of the two 82-ton expended solid rocket boosters of the space shuttle is recovered by the use of a 54-foot-diameter drogue parachute and three main parachutes, each 115 feet in diameter.

Recovery parachutes were used in the first model rockets made by Orville H. Carlisle in 1954. Since then, much has been learned about model rocket parachutes and many new techniques have been developed. A lot of serious research has been done concerning model rocket parachute recovery techniques.

A simple parachute may be added to the nose-blow model rocket or substituted for a streamer. It's surprising that a workable model rocket parachute

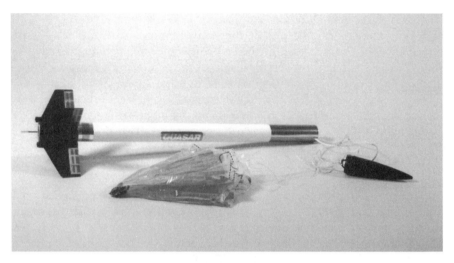

FIGURE 12-5   A model rocket with parachute recovery.

is so simple and so reliable. In fact, for many years, model rocket parachute recovery reliability was far greater than that for full-size rocketsondes and research rockets.

Model rocket parachutes are commonly made from thin polyethylene plastic sheet or film ranging in thickness from 0.00025 inch to 0.001 inch. Model rocket manufacturers sell parachute kits that include brightly colored parachute canopies of various sizes and shapes. These colored canopies are easier to see against the sky than parachutes that are homemade from transparent dry cleaner's clothing bags.

Some high-performance competition parachutes are made from very thin and very strong plastic film called Mylar. Such thin parachutes are also used when the storage space inside the model is small. Mylar film is available in aluminized form and is superior for parachute canopies because it glints in the sunlight and can be seen for very long distances even through heavy haze. Mylar parachutes have a "memory." They "remember" the flat condition in which the plastic film was originally laid down in the factory. Therefore, they tend to open up and resume this original flat condition even at very low temperatures.

Polyethylene parachutes get very stiff at temperatures below 40 degrees F. Therefore, in winter flying conditions some polyethylene parachutes fail to open. This results in a recovery device known as a plastic wad that does not have a very high drag coefficient.

Parachutes used in model rocketry are simply flat sheets of materials. Therefore, they are technically not parachutes but *parasheets*. A true parachute isn't made from a flat piece of material and cannot be laid out flat on the ground. It's made from wedge-shape gores of material that are sewn or fas-

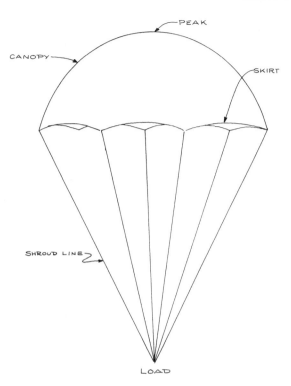

FIGURE 12-6   The parts of a parachute.

tened together to make a canopy that's hemispherical or semihemispherical in shape when inflated, as is shown in Figure 12-7.

One type of parasheet approaches a true hemispherical parachute in shape and drag characteristics. This is known as a *gathered parasheet*. It's made as shown in Figure 12-8 by taking a flat piece of material, gathering the corners, and looping each shroud line around a gathered corner.

Although model rockets actually use parasheets, it's simpler to call them parachutes or simply chutes. So that's what we'll do from now on. But now you know the difference between a real parachute and one used on a model rocket.

Shroud lines are normally made from cotton thread or carpet thread. The shroud lines are usually attached to the skirt of the canopy by strips of adhesive tape. If you're making your own parachute from a plastic dry cleaner's bag or a shopping bag, a suitable way to attach the shroud lines is with 3M's Scotch 810 Magic Transparent Tape or equivalent. For cold-weather flying, use one of the tapes designed for sealing food freezer bags. Most of these tapes will last longer than the parachute, which gets ripped, torn, worn, and probably scorched after a few dozen flights.

To prevent the shroud lines from pulling away underneath the tape, loop

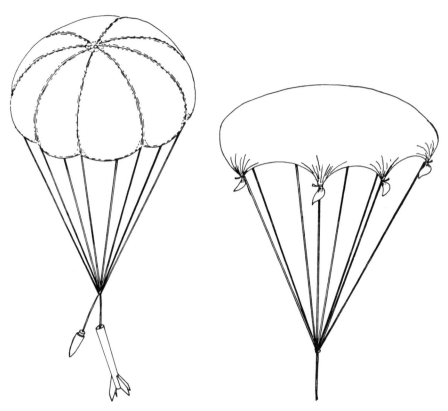

FIGURE 12-7 A true parachute is made from triangular gores of material joined to create a hemispherical canopy shape when deployed.

FIGURE 12-8 A gathered parasheet approximates the drag characteristics of a hemispherical parachute. It is made from a flat piece of material gathered where the shroud lines are attached.

them under the tape as shown in Figure 12-9. Or tie a knot in the end so that it won't pull under the tape.

A parachute more than six months old should be checked carefully before you use it. The shroud-line tapes should be replaced with new ones if the tape fails the handheld pop test where you grab the load end of the shroud lines and snap the parachute open with a quick motion of your arm. If shroud lines come loose during this test, more of them will come loose in flight. This leaves your model dangling under a parachute that acts like a streamer.

You should try to make the shroud lines all the same length because this results in a symmetrical parachute that won't slip sideways in flight and thus has a higher descent rate. Little hard data exist on the best shroud-line length for a given canopy size. My experience indicates that the shroud lines should never be shorter than the major dimension of the parachute canopy. Thus, a 12-inch parachute should have shroud lines at least 12 inches long. I have also

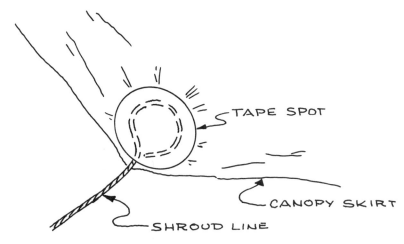

FIGURE 12-9   Shroud lines that are taped directly to the parachute canopy skirt are usually looped under the tape to prevent them from pulling out from under the tape when the parachute opens.

used shroud lines one and one-half times the major dimension of the canopy and won a few Parachute Duration contests with these.

Here's a good research project: How is the descent or sink rate of a parachute affected by the length of the shroud lines? Make several parachutes of identical size, shape, and material. Attach shroud lines of different lengths to different parachutes. Drop these from known heights indoors in calm air. The farther you can drop them, the better the data. Time their descents to the floor using a stopwatch. Do many, many drops—the more the better. This sounds like an easy project, and it is. Too many research-minded model rocketeers have become enamored by complex theoretical projects and have neglected research on numerous everyday model rocketry problems.

The desired number of shroud lines depends on the size and shape of the parachute. Parachutes have only a few basic shapes, some of which are shown in Figure 12-10. These are round, square, hexagonal (six-sided), and octagonal (eight-sided). Few circular parachutes are made and used because it's easier to cut out the hexagonal and octagonal shapes. Most chutes have a shroud line attached at each corner—six shroud lines for hexagonal chutes, eight for octagonal chutes.

Once the shroud lines have been taped to the canopy, bring the loose ends together and attach them to the base of the nose. Some model rocketeers use a fisherman's snap swivel that allows them to attach and remove parachutes for different flying conditions.

To eliminate static electricity that may prevent a polyethylene chute from opening and to unstick any shroud tape adhesive that may have gotten on the canopy, dust both sides of the canopy with talcum powder. Most model rocketeers use baby powder for this purpose; the sweet smell tends to

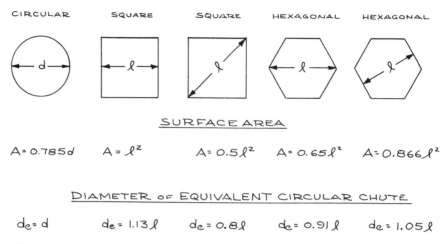

CIRCULAR   SQUARE   SQUARE   HEXAGONAL   HEXAGONAL

SURFACE AREA

$A = 0.785 d$ $A = \ell^2$ $A = 0.5\ell^2$ $A = 0.65\ell^2$ $A = 0.866\ell^2$

DIAMETER of EQUIVALENT CIRCULAR CHUTE

$d_e = d$ $d_e = 1.13\ell$ $d_e = 0.8\ell$ $d_e = 0.91\ell$ $d_e = 1.05\ell$

FIGURE 12-10 Some of the basic parachute canopy shapes tested by Kratzer and colleagues at the University of Maryland.

counteract the usual rotten-egg stink that builds up inside well-flown model rockets.

Model rocketeers have developed a wide variety of means to fold and pack parachutes into model rockets. I still use the method taught to me in 1957 by Orville H. Carlisle, the inventor of the model rocket. It has worked perfectly all these years. The Carlisle Method is shown in Figure 12-11. Follow the steps shown, and you'll have a tightly rolled parachute cylinder that will easily slide into the body tube and easily be ejected. When a chute packed by the Carlisle Method is ejected, it unrolls and deploys very quickly. I've heard chutes deploy and fill with air so quickly that they pop.

Always use flameproof recovery wadding except in models that have built-in ejection baffles and thus don't require recovery wadding. And I take one additional step to be absolutely certain that the chute doesn't get scorched or spot-burned by any ejection charge gas that happens to leak past the wadding. I wrap the rolled chute in one square of flameproof wadding. This is equivalent to the deployment bag used on many full-size parachutes. It peels off instantly, and it always works. It prevents the chute from getting ripped during the ejection process, too.

A word about wadding: *Always use flameproof wadding.* Don't use ordinary toilet paper or tissue. You do not want to have smoldering pieces of wadding land in dry grass and start a grass fire. Always use biodegradable wadding. Fiberglass from household insulation isn't biodegradable, and you'll get your fingers and hands full of microscopic glass fibers that itch like crazy and can actually be medically dangerous. Don't use cotton wadding unless it has been soaked in a flame retardant such as boric acid, then allowed to dry.

Now the always-asked question: How big a parachute should I use in my model rocket? Well, how heavy is your model? And how long do you want it to stay aloft?

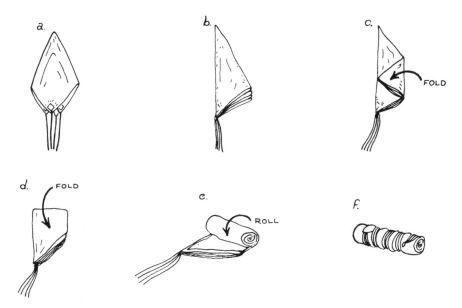

FIGURE 12-11   The Carlisle Method of folding, rolling, and packing a model rocket parachute.

Basically, the heavier the model, the bigger the parachute. And the bigger the parachute, the longer the model rocket will stay aloft. This general observation is based on the fact that a parachute produces a lot of drag. That is its function. Therefore, it obeys the drag equation

$$D = 0.5\rho V^2 C_d A$$

where A = the area of the parachute canopy.

Data from aeronautical engineering texts indicate that the drag coefficient of a parachute is about 1.5 while that of a parasheet is roughly 0.75. Although model rocketeers knew about this equation and knew it applied to parachutes as well as model rockets, we really didn't pay too much attention to what it tells us. We went by the rule of thumb: The bigger the parachute, the slower the descent—and the farther we had to chase the model on a windy day. See Figure 12-12.

Numbers were brought into the picture through a sophisticated series of parachute drop tests carried out in 1970 by Carl Kratzer, Bruce Blackistone, Greg Jones, and Larry Lyons and reported by Doug Malewicki. They made a number of timed test drops with various kinds of parachutes and standard loads from a 90-foot platform inside the Cole Fieldhouse of the University of Maryland. This provided them with a controlled environment of still air at reasonably constant temperature. The chutes used were circular, square, hexagonal, and octagonal. The first test series was made using commercially available hexagonal polyethylene parasheets 0.00075 inch thick. Sizes were 8

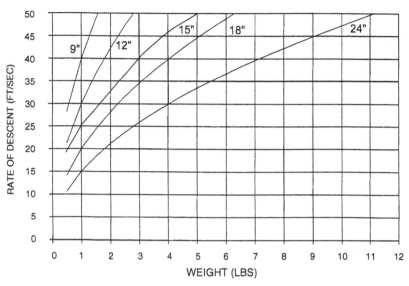

FIGURE 12-12   A descent rate chart for several typical small, round para-chute diameters. (Courtesy of Top Flight Recovery, LLC.)

inches, 12 inches, 18 inches, and 24 inches. The second series used parasheets made from half-mil (0.0005-inch-thick) transparent polyethylene from dry cleaner's clothing bags. This second series of chutes had diameters of 8 inches, 12 inches, 18 inches, 24 inches, 30 inches, 36 inches, 42 inches, and 48 inches.

A total of 240 drop tests were conducted by Kratzer with Greg Jones as the timer and Bruce Blackistone as the loyal, hardworking recovery man. The chutes were hoisted back to the ceiling platform after each drop by a high-speed parachute crane consisting of a deep-sea fishing rod, line, and reel. Each of the twenty parachutes was tested with four different payload weights, giv-ing a total of eighty different test combinations.

These tests confirmed experimentally what many model rocketeers already knew empirically from actual flying activities. However, some inter-esting new data were generated. For example:

1. The 8-inch chutes turned out to be drogue chutes—that is, they would stabilize the falling payload but did not reduce its drop speed according to the drag equations. In some cases, these small chutes didn't open fully.
2. The performances of the hexagonal and circular chutes were nearly identical.
3. Square parachutes drifted the least.

4. The second series (cleaner bag chutes) opened more easily and more completely, were approximate to the true hemispherical shape, and drifted more.

5. The square, nondrifting chutes appeared to follow the drag equation very closely and had a calculated C_d of 1.0 in neat accordance with theory.

6. However, as any competition model rocketeer will tell you, the big half-mil cleaner bag chutes performed best. Their calculated C_d was as high as 2.25. Because this is an "absurdly high value" according to Malewicki, he goes on to state, "It tells us that the chute is gliding and generating lift in addition to drag."

To see how a parachute can generate lift, look at Figure 12-13. Remember that the big half-mil cleaner bag chutes drifted the most in the drop tests. This sideways motion generated lift as shown in Figure 12-13.

This leads to some interesting speculations. Can a model rocket parachute be designed and made in the same manner as a Para-Commander or other full-scale skydiving sport parachute? Can the performance of a parachute be improved by deliberately making the shroud line lengths unequal to produce an eccentrically positioned load and therefore an induced angle of attack? Can parachute performance be improved by cutting a hole in the side of the canopy, and thus asymmetrically venting the canopy? If so, how big a hole, where should it be located, and what shape should it have for best results? What is the actual effect of a vent cut out of the peak of the parachute canopy? Does it really stabilize the parachute as some model rocketeers believe? Does it really improve the performance, and, if so, how much?

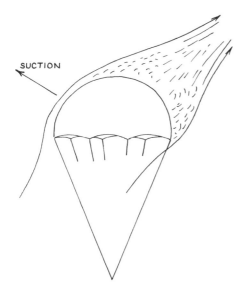

SUCTION

FIGURE 12-13 A parachute that moves sideways in the airstream will generate lift as shown.

The answers to these questions are identical: *We don't know yet because no model rocketeers have conducted and reported the results of experiments done under carefully controlled conditions such as the University of Maryland drop tests.*

It's also surprising that this sort of information isn't available in books and technical reports about full-size parachutes. What little information exists on parachute performance usually isn't applicable to model rocketry because (1) model rocket parachutes are much smaller and therefore may be greatly influenced by scaling effects, and (2) model rocket parachutes are usually made from plastic film that is nonporous while all available big parachute data relate to woven silk or nylon canopies that allow a small amount of air to bleed through them.

Therefore, a lot of model rocketry parachute research remains to be done. The experiments should be simple, inexpensive, and easily conducted. If properly designed and carried out, they could yield answers to some important questions in model rocket parachute recovery technology.

The parachute size to use can be determined by a basic rule of thumb: A parachute should have 44 square inches of canopy area for every ounce of recovered weight. The rough rule of thumb says that a 12-inch parachute is suitable for models weighing up to 2 ounces while a 24-inch parachute will handle models weighing up to 8 ounces.

Because no one has yet done any serious research and testing on the effects of the shape of the recovered model or payload dangling beneath the canopy and thereby affecting the airflow into the canopy, we often run into some unusual situations. For example, I've had some models exhibit a higher sink rate with a 24-inch chute than with an 18-inch chute.

Recalling the basic drag equation, it's easy to understand that the drag of a chute—and therefore the descent rate of any given model—increases directly as the canopy area increases (and as the square of the linear canopy dimensions). Descent speed is a direct function of the drag of a parachute. Doubling the linear dimensions should decrease the descent rate by a factor of 4.

This is an important factor to keep in mind. There are times when you *don't* want a slow descent. The launch area may be surrounded by rocket-eating trees. Or the wind may be blowing strongly. A large parachute will cause a model rocket to drift for a long distance. In contrast, when a model is carrying a fragile or heavy payload, a slow descent and soft landing may be mandatory to prevent damaging the payload during landing. Therefore, the questions of parachute size and design are subject to many compromises and trade-offs. There are no pat answers.

OTHER RECOVERY DEVICES

Featherweight recovery makes use of the familiar drag equation, too. Here, the motor casing of a very small, lightweight model rocket is ejected from the model. The model then falls very slowly because of a high area-to-weight ratio. It's literally like a feather. This recovery method is limited to very small

FIGURE 12-14   The Heli-roc kit from Apogee Components is a challenging and very high-performance autorotation recovery rocket.

and very lightweight model rockets. The NAR doesn't permit free-falling motor casings in competition, so featherweight recovery isn't seen at contests. However, once the motor is ejected high in the sky, the tiny model becomes practically impossible to see. And usually it can't be found after it lands. Featherweight recovery models usually make only one flight.

Recovery by autorotation is rarely used, although a few kits exist for those model rocketeers who want to try their hand at this. In fact, the NAR has a competition category for Helicopter Recovery Duration. Some models eject or release helicopter-like rotor blades that spin and lower the model slowly to the ground. Other models allow their tail fins to cock to one side when the ejection charge goes off, thereby causing the model to spin rapidly during descent.

And then there's *glide recovery*.

13

GLIDE RECOVERY

For decades, people have wanted to *fly* into space and *fly* back from space in winged, airplane-like vehicles. From 1925 to 1945, many different rocket airplanes were designed and flown in Germany and the Soviet Union by people who later built the first manned space craft. The Messerschmitt Me-163B Komet was a rocket-powered interceptor used by the Luftwaffe in World War II, but it was probably more dangerous to its pilots than to Allied bombers. The Soviets built, tested, and abandoned a rocket-propelled interceptor, the Berezniak-Isaev BI-2, during World War II. In the United States, high-speed aerodynamic research was conducted starting in the late 1940s with such rocket-powered manned aircraft as the Bell X-1, the Douglas D-558-2 Sky-rocket, and the North American X-15. Neil A. Armstrong, the first man to set foot on the moon, earned his astronaut's wings in the X-15 by flying it higher than 50 miles before he joined the Gemini and Apollo programs. Joe H. Engle, one of the pilots on the second space shuttle mission, also got his astronaut's wings in the X-15.

From the very beginning of model rocketry in 1957, we believed it would be possible to make a model rocket that performs like today's space shuttle—vertical takeoff under rocket power and gliding return to a horizontal landing. But we didn't fully appreciate or understand the many problems involved in this unique marriage of model airplanes and model rockets.

Rocket-propelled model airplanes have been around for a long time. The concept was first put forth by Werner von Siemens of Berlin, Germany, some-time between 1845 and 1855, but he didn't build and fly it. Probably the first were flown in Bucharest, Romania, in 1902, by Dr. Henri Marie Coanda, who in 1962 watched me launch some modern glide-recovery model rockets. (History often has a strange thread running through it.) Ron Moulton, editor of the British publication *Aeromodeller* and a longtime supporter of space mod-eling in the Federation Aeronautique Internationale, attempted to build and fly glide-recovered aeromodels using skyrocket propulsion units in England in

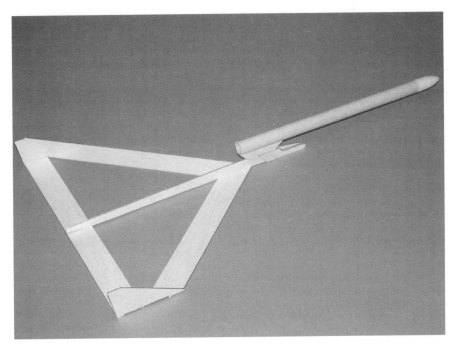

FIGURE 13-1   The Edmonds Deltie from Balsa Machining Service is one of the best beginner gliders available.

1946; his efforts were frustrated by the lack of a small, reliable model rocket motor.

Rocket-propelled model airplanes became more common in 1947 when the English Jetex rocket motors were introduced. However, Jetex motors produced very low thrust and very long duration; they were also heavier than model rocket motors with comparable total impulse. As a result, Jetex-powered model airplanes were just that—they fly under power in shallow, turning climbs with their wings always providing support against gravity. They're not normally capable of vertical takeoff.

Although model rocket motors of Type $\frac{1}{2}$A through Type B don't have any more total impulse than the Jetex motors, they do have greater thrust and lower weight. Early attempts to substitute Type A model rocket motors for Jetex motors in model airplanes resulted in some spectacular failures! The models designed for the gentle Jetex power couldn't withstand the high thrust, high acceleration, and high airspeed produced by model rocket motors. When the wings didn't peel off, the resulting violent and rapid loop produced a rather hard "prang," to use the British terminology for a flight that terminates abruptly in a high closure rate between the aircraft and the ground, usually bending the aeroplane severely.

The first publicly demonstrated glide-recovery model rocket was flown by Vernon D. Estes and John Schutz in August 1961. It was flown at the Third

FIGURE 13-2   Glide-recovery model rockets are among the most complex and advanced of all model rockets, combining the vertical takeoff of a model rocket with the gliding recovery of a wing shape.

National Model Rocket Championships (NARAM-3) in Denver, Colorado. Because the boosting portion of the model—the engine casing in this pioneer model—separated from the gliding portion, it was dubbed a *boost glider,* or simply a B/G for short. It had the normal three flight phases: vertical takeoff under rocket boost, vertical coasting flight to ejection charge activation, and recovery. But the recovery was as a glider.

The Estes-Schutz *original* boost-glider model—the very first one ever built—is now in the aeronautical collection of the National Air and Space Museum of the Smithsonian Institution in Washington, D.C. The Estes-Schutz B/G operated on a very simple principle: the motor ejection charge changed the aerodynamic configuration from that of a rocket to that of a glider. By today's standards, it wasn't a good model rocket and it wasn't a good glider. *But it worked.* Basically, it wasn't a rocket-powered glider; it was a glide-recovered model rocket. The wings were large fins at the aft end of the model, as shown in Figure 13-4. When the motor ejection charge went off, the expended motor

FIGURE 13-3 The late Dr. Henri M. Coanda (in black hat) flew the first rocket-powered glider models in Bucharest, Romania, in 1902. He visited a model rocket launch in 1962 where the author (left) and A. W. Guill (right) explained modern glide-recovery techniques. (Photo by Lindsay Audin.)

casing was ejected from the model. This not only reduced the weight of the model but triggered a mechanical latch that released control surfaces on the trailing edges of the wings/fins, permitting little springs to force these elevons up. Technically, these elevons provided negative camber to convert the fins into wings. The model then entered a glide.

Once Estes and Schutz demonstrated a workable glide-recovery model, the field of B/Gs literally exploded into creative development. Other model rocketeers, including many who couldn't get a B/G to work at all before, proceeded to get boost-glider models into the air and working. This is an interesting phenomenon that often occurs in technology. Something may be considered impossible or extremely difficult until someone finally gets it to work, no matter how crude the workable solution may be. Once the breakthrough happens and people *know* it can be done because *somebody else did it,* developments follow very rapidly.

In the next five years, the skies were filled with wildly developed cut-and-try B/G configurations. Estes Industries, Inc., produced the Astron Space Plane B/G kit, which worked in a similar fashion to the original Estes-Schutz B/G. However, the Astron Space Plane was designed so the average model rocketeer could build it and get it to work. The Astron Space Plane was a flying wing configuration that is rarely seen four decades later, because better B/G designs

FIGURE 13-4   The first model rocket boost glider was designed and built by Vernon D. Estes and John Schutz in 1961. This is a photo of the original model now in the national aeronautical collection of the Smithsonian Institution.

have come along. Centuri Engineering Company soon followed with Lee Piester's Aero-Bat, which is still an excellent B/G design, although the kit is no longer made. Paul C. Hans worked the bugs out of the canard configuration. Hunt Evans Johnsen of MIT built the first swing-wing variable-geometry B/G that looked like the F-111. Captain Bill Barnitz of the U.S. Air Force, an F-4 pilot, tackled the flex-wing configuration. All of us were reinventing the wheel, however, because we really had very little knowledge about why or how. We were engaged in empirical experimentation with little background in the theory of lifting surfaces and aircraft aerodynamics. This was due to another common phenomenon in technology that can and should be avoided.

In spite of the fact that model rocketry owes a great deal to model aviation, people don't come into the hobby of model rocketry by way of model aviation. They usually come right into model rocketry without having previously built a flying model of any type, not even a simple balsa chuck glider. For many years, model rocketeers and model aviators didn't—and still don't—talk to one another. The model rocketeers think the model aviators are hopelessly old-fashioned with their propeller-driven flying machines. On the other hand, model aviators consider model rocketeers to be hopeless pyromaniacs who indulge a fascination with fire and smoke while cluttering up the air with ugly little plastic parachutes.

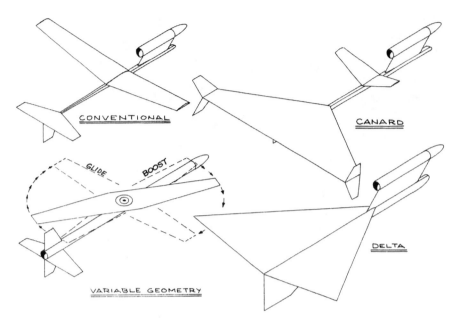

CONVENTIONAL

CANARD

GLIDE   BOOST

DELTA

VARIABLE GEOMETRY

FIGURE 13-5   Some of the basic glide-recovery model rocket types.

This lack of communication forced model rocketeers to go through a long period of trial-and-error experimentation with B/G models, making the same mistakes that the model aviators had experienced years before. Model aviators could have explained a great deal about hand-launched and catapult-launched gliders to model rocketeers, because a B/G turns out to be a rocket-launched version of a hand-launched glider in most cases.

Whenever communications between model rocketeers and model aviators were established, even briefly and almost accidentally, the results were spectacular. This should serve as a lesson to us all. A lack of communication—an inability or unwillingness to listen and a failure to seek all pertinent information—can lead to a lack of progress in any field of human endeavor, not just model rocketry.

Thus, the second big breakthrough in B/G technology came about as a result of communication between model rocketry and model aviation. Larry Renger, a model aviator and then a senior at MIT in aeronautical engineering, took a basic hand-launched glider and put a model rocket motor up front, mounted on a pylon. It was a conventional wings-and-tail airplane configuration, something none of us had bothered to try, because our early experiments had failed for many reasons. With the model rocket motor up front, the model was very nose-heavy and became essentially a ballistic vehicle like a model rocket. When the ejection charge popped the motor casing out of the model, it removed weight from the front and shifted the CG rearward, reducing the gliding weight of the model at the same time. The glider then settled

FIGURE 13-6   The original front-motored rocket-boosted glider design, the Renger Sky Slash, as built from Estes Industries plans.

into a nice low-sink-rate glide. This was the famous Renger Sky Slash design— a rocket-powered glider, not a gliding model rocket. The emphasis was on the boost *glide* rather than *boost* glide, as in previous B/G designs.

Properly trimmed, a Renger Sky Slash would outglide almost any other type of B/G then flying, even though the original Renger design had been simplified and changed in the plans that Estes Industries published. The Sky Slash design dominated boost-glider duration competition here and abroad for over a decade. Then it was challenged by the developments in flex-wing technology revealed by the Bulgarians at the 1980 World Spacemodeling Championships held at Lake-hurst, New Jersey, where the Bulgarian B/G team walked away with the gold team medal plus gold and bronze medals for individual first and third places.

The ultimate refinement of the front-motored conventional configuration B/G came with the development of the pop pod to eliminate the potential safety problems associated with ejecting an empty motor casing. There is some question about who actually invented the pop pod. I was flying B/G models with pop pods in April 1965. I subsequently learned that Larry Renger had designed the FlaminGo with a pop pod at about the same time. Here is an example of parallel technological development because it was obvious that something like a pop pod would be required, in order to conform to the com-petition rules. Besides, a pop pod would greatly improve the glide perform-ance of a B/G if you could jettison *everything* that contributed weight or drag to the glider portion of the model.

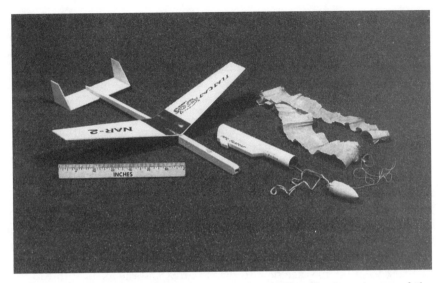

FIGURE 13-7   A boost glider with a pop pod. The ejection charge of the model rocket motor pops off the pod nose and ejects the pod streamer. Reaction forces the pod backward, disengaging it from the glider portion. This is the first version of the author's popular Flat Cat beginners' B/G model.

The boost-glider concept permits the model to drop its propulsion portion just as the space shuttle drops its SRBs and External Tank. But even today, professional rocket engineers are at work on future single-stage-to-orbit (SSTO) designs that don't drop *anything* and use up only rocket propellants during the flight.

About 1971, model rocketeers also began to think about the challenge of glide-recovery models that didn't drop *anything* in flight. The result was the *rocket glider,* the RG. It differs from the B/G in that the B/G is permitted to jettison its propulsion portion while an RG must go up like a rocket and glide back without separating *anything* in flight.

A rocket glider is truly a glide-recovered model rocket. It follows the basic single-stage model rocket concept: what goes up together must come down together in a safe manner. The 1970s saw the long, painful development of RG technology just as the 1960s saw that of the B/G.

An RG model rocket is exceedingly difficult to build and fly, making it probably the most advanced area of model rocketry. There are three ways to make an RG work:

1. Use the ejection charge to cause a change of the CG of the model. This can be accomplished by shifting the position of the motor casing at the time of ejection charge activation, moving the motor rearward, catching it, and holding it there.

2. Use the ejection charge to change the aerodynamic configuration of the model. The ejection charge can be used to activate elevons or other surfaces, as in the early B/G models, or to actually change the relative positions of the aerodynamic surfaces of the model.

3. Use the ejection charge to cause a change that combines both of the above.

4. Utilize radio control that permits the builder and flyer to keep the RG in a vertical climb during boost and make the transition to glide phase by controlled means.

Howard Kuhn, many times U.S. senior national champion, methodically considered, designed, built, and tested dozens of RG designs to evaluate the first three of the above solutions. He came to the conclusion that solution no. 2 was the best. His design, the Buzzard, featured a *movable wing* that slid back and forth on the fuselage. The Buzzard looked like a standard, conventional configuration hand-launched glider or B/G. During boost, the wing was held at the rear end of the fuselage next to the elevators and rudder by a thread that went forward and passed through the forward power pod body in front of the motor. The Buzzard was launched like a model rocket, and, with the wing all the way back at the end of the fuselage, it was like an ordinary model rocket with a pair of big fins on it. When the ejection charge went off, the thread holding the wings was burned through. A rubber band pulled the wings forward on the fuselage up against a forward stop and the Buzzard converted to a glider.

Careful design work by experienced model rocketeers such as Chris Flanagan and Jim Pommert perfected RG design based on solution no. 1 with RG designs of conventional configuration, having moving pods or moving motors to shift the CG location, or, much more difficult, shifting weight merely by the expenditure of propellant in the model rocket motor. This last design approach seems to result in models that start to level off and glide at or shortly after burnout, as you might expect.

The Estes T25 Centurian is one of the only commercially available radio-controlled rocket-glider kits available. It has a twin-tail design and can perform aerobatic stunts after the boost phase of flight. It represents the frontier of model rocketry glide recovery today. It uses a special model rocket motor with no time delay or ejection.

RG is still an advanced area of model rocketry. In spite of Kuhn's outstanding empirical work in RG that is similar to Vernon Estes's development efforts on the Astron Space Plane B/G kit in 1961, RG has not yet received the relentless scientific scrutiny given to the B/G field.

In fact, the whole field of glide recovery is one of the hottest in model rocketry. An entire book could now be devoted to glide recovery alone. Since we can't possibly cover all the details of this complex field, this chapter will have to serve primarily as an introduction to one of the most fascinating areas of model rocketry.

Much of the following information applies to both boost gliders and rocket gliders. Some is pertinent only to boost gliders because they've been around longer and we understand them better.

FIGURE 13-8   Regarded as the world's foremost RCRG expert, Phil Barnes preps his S8E/P RCRG while David O'Brian looks on at the 14th World Space-modeling Championships in the Czeck Republic in October 2002.

FLIGHT OF GLIDE-RECOVERY MODEL ROCKETS

Boost gliders and rocket gliders may be considered either as model rockets with gliding recovery or as gliders with rocket boost. Depending on where you place the emphasis, they are *boost* gliders or boost *gliders*. The same holds true for rocket gliders.

The flight of a glide-recovery model rocket can be generally divided into two separate and distinct phases: the boost phase and the glide phase. The requirements for stability, structural integrity, and aerodynamics differ markedly from the boost phase to the glide phase. This simple fact means that a glide-recovery model rocket must always be a compromise between a model rocket and a glider. The sorts of technical trade-offs made by the designer dictate which kind of glide-recovery model rocket the result will be.

The same simple fact also dictates that a change must take place for the model to make a successful transition between boost and glide phases. In the boost glider, the booster airframe and motor casing are jettisoned, causing a change in the CG location, gliding weight, and other factors. In a rocket glider, the ejection charge of the motor must shift the CG location or change the aerodynamic configuration but there is no change in weight—unless the RG is radio controlled.

Boost phase is that portion of the flight during which the model performs like a ballistic vehicle—a model rocket—launched from a standard launch pad with electrical ignition. It's propelled aloft in a near-vertical trajectory by the high thrust of the model rocket motor. It trades vertical velocity and momentum for altitude in the time-honored fashion of a model rocket during the operation of the motor's time delay. During boost phase, a glide-recovery model is not supported by lifting surfaces operating against gravity. Whatever aerodynamic lifting surfaces are exposed to the airstream during boost either must have no effect on the ballistic flight path of the model or must stabilize the model in the same manner that fins stabilize an ordinary model rocket.

The characteristics of boost phase include high acceleration and high airspeeds. These can combine to produce a situation aptly described by the model rocket term introduced by the Czechs: striptease. This is an understandable word in many languages! Large wings rip off and poorly made glue joints may turn loose. Therefore, glide-recovery models must be strong and very good construction techniques must be used. The models usually have short, stubby wings in comparison to nonboosted high-performance models or full-size sailplanes.

The glide phase has entirely different characteristics. The model in a glide is supported by the lifting force of its wings, which sustain it against gravity by the forward airspeed creating airflow over the wings. This airspeed is as low as 10 feet per second or even less. The sink rate, or vertical descent rate, may be as low as 1 foot per second. Since the sink rate of a glider is directly related to the ratio between the lift and drag of the glider, very high lift-drag ratios on the order of 10:1 or more are desirable if you can obtain them.

As during the boost phase, the gliding model must be stable in pitch, yaw, and roll. But the pitch and yaw axes can no longer be considered identical because most glide-recovery models are not symmetrical in the pitch and yaw axes. The glide-recovery model must have very high inherent stability because winds, gusts, and thermals (rising air columns and bubbles) will continually disturb its equilibrium during glide. If it pitches nose-up, the forces generated by its lifting/stabilizing surfaces must bring the nose down into an equilibrium or balanced gliding condition again. The same holds true for nose-down pitching. If the model rolls, it must right itself. If it yaws right or left, it must correct itself through proper design. Unlike an ordinary model rocket, a glide-recovery model during glide must always "know" which way is up—that is, where the gravity vector force is.

A fin-stabilized model rocket is not a good glider and a glider is not a good model rocket. Since the two flight phases are very different in their requirements for design and construction, compromises must be made. If maximum flight duration is the objective, this can be obtained in either of two ways. A modeler can strive for maximum altitude during boost, compromising good glide characteristics and obtaining long flight duration by simply having the model take a long time to descend from a very high altitude in spite of the fact that it may have a high sink rate. Or the glide-recovery model designer can work to achieve maximum performance during the glide phase,

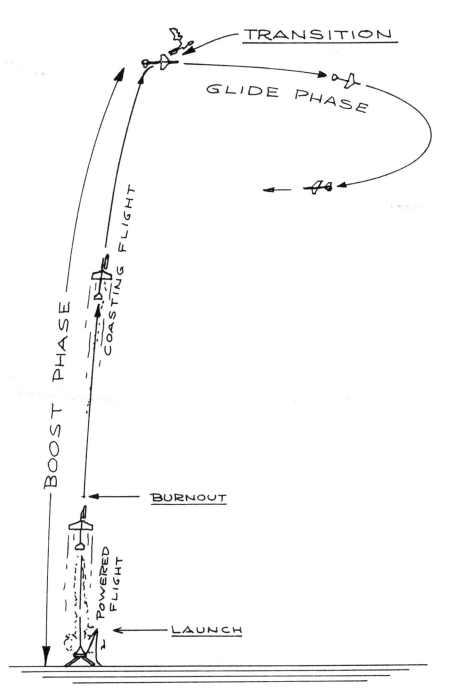

TRANSITION

GLIDE PHASE

COASTING FLIGHT

BOOST PHASE

BURNOUT

POWERED FLIGHT

LAUNCH

FIGURE 13-9   The basic flight regimes of a glide-recovery model rocket.

cutting back on the altitude gained during boost in favor of low sink rate and other more favorable glide characteristics.

Performance depends a great deal on the weather at the flying site when you get ready to launch. Many modelers take two different glide-recovery models to a contest: a foul-weather model and a fair-weather model. A foul-weather model usually has very good boost characteristics and a very stable glide, while a fair-weather model is usually designed for absolute maximum glide performance in calm or nearly calm air.

STABILITY AND CONTROL OF GLIDE-RECOVERY MODELS

Except for the radio-controlled glide-recovery model rockets, small glide-recovery model rockets of the sort built by most modelers can't be controlled during their flight. They're too small to carry even a modern solid-state radio-control receiver, batteries, and servos. Therefore, a glide-recovery model rocket must have inherently good stability and be able to maintain its equilibrium during flight because of this good stability.

Roll stability during boost may not be required. In fact, some models are deliberately rolled slowly during boost by means of offset pods or motor thrust lines. This causes the model to "screw" its way up into the sky. Basically, this averages out any small pitch or yaw motions that occur during boost phase and is considered a quick and dirty fix for any model rocket—glide recovery or not—that may be *slightly* unstable or dynamically undamped. Too much roll during boost phase can be self-defeating, however, because it can greatly reduce the peak altitude attained.

During glide phase, roll stability can be obtained by using a dihedral angle on the wings as shown in Figure 13-10. This dihedral angle should be between 10 and 20 degrees per wing. It can be made up of compound dihedral angles—that is, a straight center section with the tips turned up to the same location they'd have if the wing had straight dihedral. (This was the quick fix used by aeronautical engineers on the F-4 Phantom II jet fighter plane; it needed more dihedral for roll stability, but adding straight dihedral to its wings would have required extensive redesign of the wing-fuselage structure.) The optimum dihedral for a wing is actually elliptical, as shown in Figure 13-11.

Stability in the yaw axis is obtained by a combination of vertical stabilizers or stabs (which may have movable portions called rudders), wing sweep-back, and/or wing dihedral. Normally, adequate yaw stability can be obtained with a vertical stab area equal to 5 to 10 percent of the wing area, as we'll see later. The biggest control and stability problem for all glide-recovery models in both boost and glide phases is in the pitch axis. See Figure 13-12.

During boost phase, an improperly designed, constructed, or balanced glide-recovery model—except the flex-wing types that store the gliding wings folded inside a body tube or the parasite glider types that are very small in comparison to the boost pod or vehicle—will pitch up into a loop or pitch down into an outside loop. Although these aerobatics are often spectacular, the model usually prangs. Or it loops so tightly that it doesn't gain sufficient

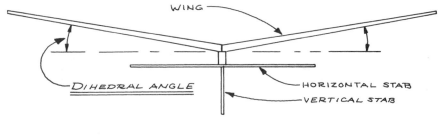

FIGURE 13-10   The dihedral angle of a B/G's wings as seen from the front.

altitude for a long glide. Or the range safety officer or judge yells, "DQ!" That means the flight's been disqualified as unstable and/or unsafe.

The radius of the loop caused by pitching depends on (1) the location of the CG and (2) the wing loading of the model (in ounces per square inch of wing area, for example). The farther aft the CG or the lower the wing loading, the smaller the loop radius. Therefore, the most desirable model characteristics during boost phase are *forward location of the CG and high wing loading.*

The CG characteristic is why a front-motored glide-recovery model performs better than a rear-motored one. The wing loading factor is the reason why a parasite glider—a gliding portion of a B/G that's hung on the side of a much larger booster vehicle (such as the Quest Aurora) and therefore has very high wing loading when attached to this booster—works so well during boost.

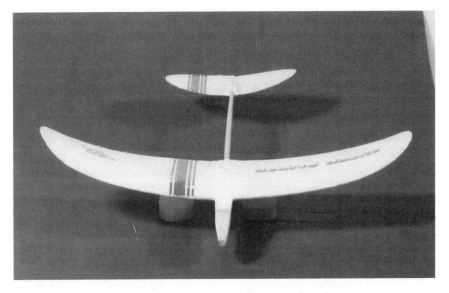

FIGURE 13-11   Elliptical dihedral of a wing.

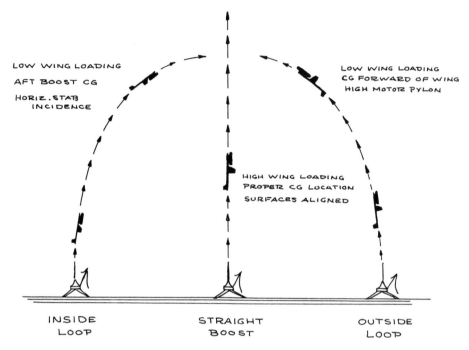

LOW WING LOADING
AFT BOOST CG
HORIZ. STAB
   INCIDENCE

LOW WING LOADING
CG FORWARD OF WING
HIGH MOTOR PYLON

HIGH WING LOADING
PROPER CG LOCATION
SURFACES ALIGNED

INSIDE
LOOP

STRAIGHT
BOOST

OUTSIDE
LOOP

FIGURE 13-12   Pitch axis stability problems of glide-recovery model rockets during boost phase.

Improper balance or trim (location of the CG) during glide phase will cause the glide-recovery model to pitch up into a stall or dive into the ground. A badly trimmed glider may also get into increasingly steeper and more violent stalls, ending up in a spiral dive to the ground. Stalls are caused by the CG being too far aft—a tail-heavy model, in other words. At its worst, this aft CG condition leads to sharper and deeper stalls until the model goes into a spiral dive from which it never recovers. With a nose-heavy model where the CG is too far forward, there's very little gliding flight at all. The model simply goes into a dive and prangs.

Control and stability of a glide-recovery model during the boost and glide phases are a matter of *balancing* the forces generated by the various surfaces lying in the pitch plane—the wings and the horizontal stabilizer (if any).

As shown in Figure 13-13, the CP of a flat plate or symmetrical airfoil, including symmetrical model rocket fins, is always located at a distance of 25 percent aft of the leading edge of the average chord of the surface. Now, what do we mean? Figure 13-14 shows you. The *chord* of a wing, stab, or fin is its fore-and-aft dimension parallel to the body tube or fuselage. Unless a wing has a rectangular planform (a "Hershey-bar wing") with the same chord from root to tip, the average chord is the sum of the root chord plus the tip chord, the sum being divided by 2. You can precisely locate the average chord by geometrical construction as shown in Figure 13-15.

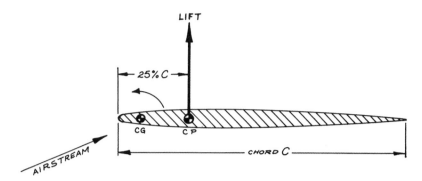

FIGURE 13-13   Cross-section of a symmetrical airfoil with the CP located at 25 percent of the chord and the CG located ahead of the CP. Pitching moment will always be in the direction of the oncoming airstream, making the airfoil always point into the airstream. This is a stable condition for boost phase but will not produce a glide.

If we place the CG ahead of the CP and put the airfoil in a glide condition at an angle of attack as shown in Figure 13-13, the lift force concentrated at the CP will tend to make the wing rotate and pitch down or into the airstream, reducing the angle of attack to zero. The surface always seeks zero angle of attack. This is precisely the situation we want for the fins of a good fin-stabilized model rocket but not for a glide-recovery model rocket during glide phase. This condition will simply make it dive right into the ground.

Therefore, to make a model glide with a CG forward of the wing's CP, we must have some mechanism that will bring the nose up when the model starts to dive and, conversely, bring the nose down when the model starts to climb and approach a stall. We must somehow obtain a *balance of forces*. Our glide-recovery model is just like a seesaw with its balance point being the CG. Anything that makes it want to dive should immediately produce a force to make it climb and bring it back to level flight again. And, conversely, anything that makes it want to climb should produce a pitch-down force that will bring the nose back to level flight.

The model must be stable in the pitch axis just as it's stable in the roll axis, thanks to wing dihedral, and in the yaw axis, thanks to the vertical stab, wing sweep, and wing dihedral. Any tendency of the model to wander away from its balanced, stable, level gliding flight condition must be counteracted by forces that will return it to this balanced gliding condition.

As you can see, this stable glide condition is radically different from the stability condition required during boost phase. Therefore, something must happen to the model to convert it from a ballistic, nose-heavy, fin-stabilized model into a gliding model. And this should logically occur at or near peak altitude achieved during the boost phase.

A glide-recovery model is designed and built so that the motor ejection charge action changes the model to convert it from a rocket to a glider. This is

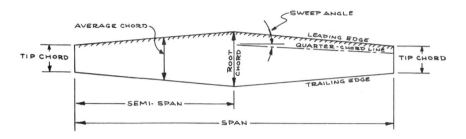

FIGURE 13-14   Definitions of the dimensions of a wing or fin.

usually accomplished by either changing the location of the model's CG or changing the model's physical characteristics, or a combination of both.

The Estes-Schutz B/G reduced its weight by motor ejection (now no longer permitted), thereby reducing its wing loading. The ejection of the motor casing released a complex mechanical mechanism that permitted control surfaces—elevons—on the trailing edges of the wings to be deflected upward. From Figure 13-16, you can see that this produced a pitch-up force to oppose the pitch-down force of the wing lift.

Nearly all rear-motored B/G models use the ejection of a propulsion pod or module to release control surfaces. However, these pods or modules must descend with a streamer or parachute attached because the NAR competition rules (and common sense) don't allow the ejection of an empty motor casing

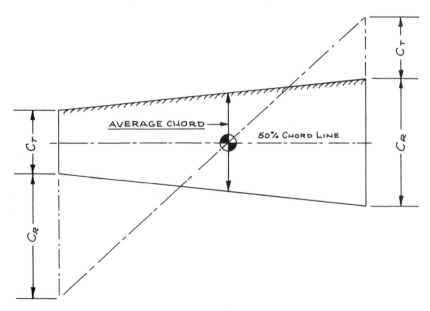

FIGURE 13-15   How to locate the average chord by geometrical means.

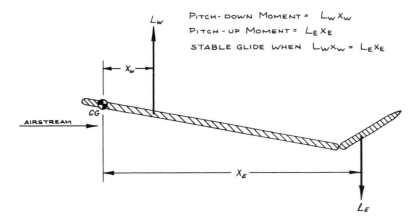

FIGURE 13-16   Use of a trailing-edge control surface on a flying wing to provide pitch stability. The wing alone gives a pitch-down moment; the control surface produces a pitch-up moment. When the two pitching moments are equal, the wing will glide.

from a model rocket. (Not all model rocket motor casings are small, not all are made from paper, and all *are* difficult to see when falling free.)

Rear-motored B/G models usually have delta-wing (or its derivative, the flying-wing), canard (horizontal stabilizer in front), or variable-geometry (swing-wing or scissors-wing) configurations. However, the simplest of all B/G models to design and build—but not to trim and fly—is the front-motored conventional configuration rocket-boosted glider pioneered by Larry Renger.

With movable-surface glide-recovery models, we saw how the pitch-down lift force of the wing with a forward CG was counterbalanced by the lift force generated by control surfaces to maintain a stable glide. But a front-motored pop-podded B/G model and a front-motored, CG-shifting RG model usually don't change any aerodynamic surfaces when the motor ejection charge activates. Instead, the CG of the model is shifted, and, with the B/G model, the wing loading is reduced by the jettisoning of the power pod or the separation of the booster rocket in the case of the parasite B/G types. Thus, in a CG-shift model, no new lifting forces are created at the transition from boost to glide, but the *relationships* between the lifting forces and the CG are changed by changing the CG location. Basically, we change the pivot point of our aerial seesaw.

Figure 13-17 is a simplified side-view representation of a CG-shift glider with all elements germane to this discussion shown. You can imagine that the vertical stab and other elements are there if it will make you feel better. The wing is located on the forward portion of the model and a horizontal stabilizer is located at the aft end of the model.

During boost phase, the CG is located at or in front of the leading edge of the wing. The model, therefore, acts like a fin-stabilized model rocket because

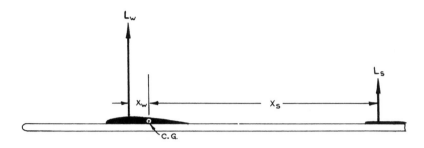

$$\text{PITCH-UP MOMENT} = L_w X_w$$
$$\text{PITCH-DOWN MOMENT} = L_s X_s$$
$$\text{STABLE GLIDE WHEN } L_w X_w = L_s X_s$$

FIGURE 13-17   Graphic representation of a CG-shift boost glider in gliding condition with the CG shifted to 50 percent of the wing chord. The wing now has a pitch-up moment that is balanced by the pitch-down moment of the aft-mounted horizontal stabilizer. For clarity, the rudder is not shown.

the CPs of all surfaces are behind the model's CG. If the CG was not shifted backward at transition, the model would continue to behave like a fin-stabilized model rocket and fly its parabolic trajectory right back into the ground. But if the motor ejection charge jettisons a front-mounted pop pod containing the expended motor casing (B/G model) or moves the power pod rearward (RG model) at transition, a totally different aerodynamic situation is created.

Let's suppose that the CG is shifted aft to a location that is 50 percent of the wing chord as shown in Figure 13-17. The CG is now *behind* the CP of the wing and *ahead* of the CP of the horizontal stab.

If the wing were up there all alone, the wing lift concentrated at the wing CP would cause the wing to pitch up. You can create this situation by taking a piece of $\frac{1}{16}$-inch sheet balsa or heavy cardboard about 3 inches wide and about 12 inches long. Try to make it glide by itself. You'll find that its leading edge will pitch up and the whole sheet will descend to the floor in a rotational flip, rotating about the spanwise axis at 50 percent of the chord.

To counteract this wing pitch-up force, a counteracting pitch-down force must be created to keep the model in level flight. This is done by adding the horizontal stab at the rear end of the fuselage. The horizontal stab is really just another little wing. It also produces lift, but its CP is a long distance behind the model's CG. This lifting force of the horizontal stab produces the necessary pitch-down force.

Although the lift force of the wing is much greater than that of the stab because of the greater area of the wing, the lifting force of the wing is closer to the model's CG. Therefore, the lesser pitch-down force produced by the smaller stab balances the wing's pitch-up lifting force. It's like balancing a 50-pound boy with a 250-pound middle linebacker on a seesaw. You can do it if you put the 250-pound linebacker closer to the pivot point than the 50-pound boy. This is because their *moments* will be equal.

What is a *moment?* Simply the product of the force and the distance through which it acts. It is the weight times the distance from the balance point or the force times the distance from the CG. As you can see in Figure 13-17, the pitch-down moment of the stab (a small force acting over a long distance) is equal to the pitch-up moment of the wing (a large force acting over a short distance).

If the model starts to dive, the wing builds up more lift force as a result of increased airspeed than the little stab. Thus, the forces become unbalanced in the pitch-up direction. The model pitches up to its balanced glide angle. If the model starts to climb, the wing loses lift faster than the stab and the pitch-down moment of the stab forces the nose down.

Yes, it's possible to have the wing and the stab exactly the same size. This is called a *tandem wing* model. But the glide CG point must therefore be about halfway between the two tandem wings. The CG shift from boost to glide will have to be even greater with this tandem wing configuration, won't it?

The *canard* configuration can be looked at as a conventional model with a very small wing up front and a very large horizontal stab in back. The horizontal stab at the rear then contributes most of the lifting force of the total model and it becomes a wing. Therefore, the CG shift for a canard configuration is even greater than that of a tandem wing and much greater than a conventional configuration.

This is a highly simplified explanation of glide structures and aerodynamics. I had to digest a lot of data from technical reports and books before boiling it down to this point. But you don't have to be an aeronautical engineer to design a good glide-recovery model. Nor do you have to be an expert to build and fly one. You do have to understand what we've discussed thus far and you do have to be careful in your workmanship. And you should take your first steps in glide-recovery model rocketry with boost gliders rather than the more complex rocket gliders. This is a very complex area of model rocketry.

BASIC DESIGN RULES FOR GLIDE-RECOVERY MODELS

To design a CG-shift glide-recovery model from scratch, you should have access to airfoil data such as in *Theory of Wing Sections* by Abbott and Von Doenhoff (see the Bibliography). The method I used to make my calculations can be found in Frank Zaic's excellent and informative book *Circular Airflow and Model Aircraft*, also listed in the Bibliography.

However, you don't need to delve into these sources unless you want to, because I've worked out the following set of empirical design rules for front-motored CG-shift B/G or RG models. These rules aren't arbitrary. I didn't develop them by cut-and-try methods. I researched the literature, did a lot of design work in trade-offs and such, built models from the research results, tested the models in flight, then came back and went around the circle again to refine the rules. They work for me and they were put to the test by other

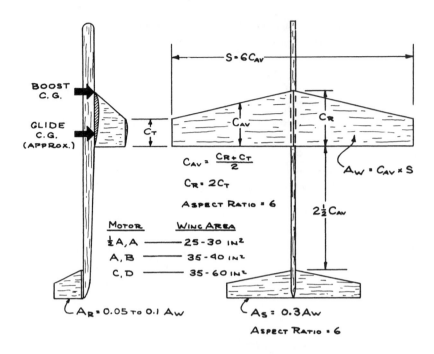

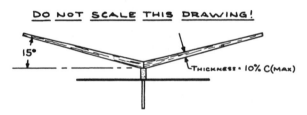

FIGURE 13-18   Stine's Basic B/G Design Rules.

model rocketeers, young and old, tyro and expert. They work. They'll work for you, too, if you follow them.

The basic Stine Design Rules are shown in Figure 13-18. This is a simplified three-view drawing of a hypothetical front-motored conventional configuration without its pop pod. It's not drawn to scale but for maximum legibility.

Start the design with the basic dimension: the average wing chord. The wing span should be five to eight times the average chord dimension. This will give your glider wing an *aspect ratio* of 5:1 to 8:1, as defined in Figure 13-19. Although high aspect ratios of 10:1 or more are used on high-performance model and full-size sailplanes, it's very difficult to build a high-aspect-ratio B/G or RG wing that will stay together during the high-acceleration, high-airspeed boost phase of flight. Only a few-percent increase in performance

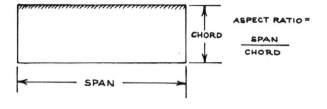

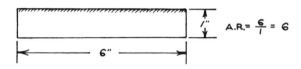

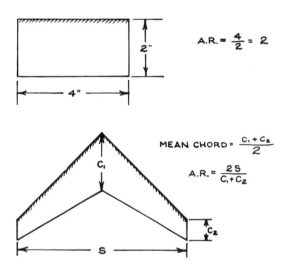

FIGURE 13-19   How to find the aspect ratio of a wing or fin.

and decrease in wing-induced drag can be obtained with aspect ratios higher than 8:1.

For the best glide performance, a wing with a sweepback of 20 degrees or less—preferably none—should be used. High sweepback angles create high induced drag. *Moderate* sweepback angles up to 20 degrees may increase the model's yaw stability, but good yaw stability can be obtained by a properly sized vertical stab.

Although an elliptical wing shape, such as that used on the legendary Supermarine Spitfire, is the optimum aerodynamic wing shape for lowest induced drag and best stall characteristics, this optimum elliptical shape can be approximated within 2 percent using a straight taper wing as shown. The taper wing is similar to that used on an equally legendary airplane, the North American P-51 Mustang fighter plane. The benefits of an elliptical planform can be obtained with a taper wing where the tip chord is one-half of the root chord as shown.

Although a glide-recovery model will fly with an unstreamlined, un-shaped, unairfoiled slab of sheet balsa for a wing, it will fly better and be somewhat easier to trim if an airfoil is used. This is because an airfoil with a rounded leading edge and a tapered trailing edge will produce a higher lift-to-drag ratio than a flat plate.

And it isn't really necessary to use the classic model airplane airfoil that's curved on top and flat on the bottom. A symmetrical airfoil, such as used on fins, will glide just as well (and perhaps better since its CP remains at or near the 25 percent chord point, whereas the CP of a flat-bottomed or *cambered* classic airfoil moves *forward* with increasing angle of attack, introducing another variable into an already complex system).

The thickness of the wing should be between 5 and 15 percent of the wing chord. The maximum thickness of the airfoil should occur at 30 to 40 percent of the chord. If the wing is tapered in planform, it should also be tapered in thickness to preserve the same airfoil all along the wing.

Many model rocketeers and nearly all hand-launched glider experts believe that the thinner the wing airfoil, the better. This doesn't jibe with wing theory or with actual field data. The thicker the airfoil, the greater the lift force per unit area up to a thickness-to-chord ratio of 15 percent for most air-foil shapes. A thicker wing is a stronger wing, although a thicker wing will offer higher drag than a thin wing.

A hand-launched glider has a propulsion phase of flight about 4 feet long—the distance through which your arm moves as you heave the glider. Any glide-recovery model rocket has a propulsion phase under rocket boost that's perhaps several hundred feet long. Although a fat-wing B/G or RG may gain less altitude during boost because of the higher drag of its thick wing, this can often be more than offset by the improved glide characteristics and the higher strength of the thicker wing. One of the biggest problems with thin-wing B/Gs and RGs is their lack of strength in comparison with thick-wing models, which results in a higher risk of "striptease" or loss of the wing dur-ing boost.

Championship full-size sailplanes have thick wings in comparison to clas-sic hand-launched gliders of model aviation. So do full-size airplanes. I once owned a manned boost glider: a Piper Cherokee 140B. It had a wing with 15 percent thickness at 40 percent of the chord, an aspect ratio of 5.71:1, and a glide ratio and therefore a lift/drag ratio of 10.32:1. This is comparable to most hand-launched gliders and glide-recovery model rockets. As a matter of fact, I checked out a lot of glide-recovery dynamic stability hypotheses using my Cherokee because it offered the advantage of having the observer aboard rather than watching from the ground. It was also controllable.

The horizontal stab of a glide-recovery model should have an area of 30 percent of the wing area, plus or minus 5 percent. It shouldn't have more than 2 to 4 degrees of negative incidence or downward tilt with respect to the wing. You'll find 5 to 10 degrees of negative incidence in most inexpensive chuck gliders you can buy in a hobby shop. This negative incidence was put in to ensure maximum pitch stability when flown by anybody.

A zero-zero glider—one with the wing and the stab at the same incidence as shown in all the examples illustrated thus far—will glide fine. However, thin-wing zero-zero designs may have a tendency to get into a dive and refuse to pull out, particularly if the transition from boost to glide occurs with the model in a nose-down attitude. A very slight 2- to 4-degree negative incidence will prevent this from happening but also requires a higher wing loading and a more forward CG location during boost to keep it from looping under power. A thicker wing is a better solution because this creates more wing downwash on the stab.

Important: For zero-zero models, put the stab behind and in the wake of the wing to take advantage of the fact that a lifting wing deflects the airstream downward behind it, creating what is known as downwash. The wing down-wash on the stab makes the stab think it is flying at a negative angle of attack, and greater wing downwash eliminates the need for negative stab incidence.

The other design characteristics and dimensional relationships of the Stine Design Rules are exactly as shown in Figure 13-18. They're somewhat forgiving because you can change them 10 percent or so either way without getting into too much trouble with the design.

Note that the single vertical stab is mounted *underneath* the horizontal stab. Not only does this keep the vertical stab completely out of line of the motor exhaust jet from a front-mounted power pod but its location underneath aids the glider in rolling into horizontal flight at the top of its climb during transition. A top-mounted vertical stab operates in the wake of the horizontal stab at or near the stall point, thereby reducing yaw stability, because a properly trimmed glider operates *just above* the stall point. A top-mounted vertical stab also doesn't contribute to roll stability but works against it. These Design Rules will also produce a very fine chuck glider for hand launching. When completed, the basic glider should have a weight of about 0.025 ounce (0.71 gram) per square inch of wing area.

For best performance, a B/G or RG should be designed for the motor type it will use. For B/G models, the following basic sizes have been found to be successful:

Motor Type	Wing Area (sq. in.)
Type A	20–35
Type B	30–45
Type C	30–60

We haven't developed any sizing rules for larger B/G models yet. And there are no basic sizing rules for RG models at this time.

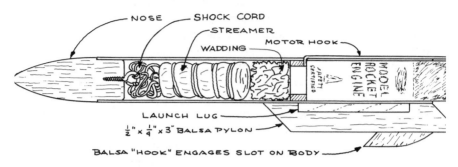

FIGURE 13-20   Cross-section drawing of a typical pop pod.

BOOST-GLIDER POP POD DESIGN

Although rocket glider models retain their motor casing and propulsion modules during glide, nearly all competition boost-glider models use a pop pod. A typical pop pod is shown in Figure 13-20.

A pop pod separates the entire weight of the propulsion unit from the model at transition, leaving only the basic glider portion of the model to glide. Anything that doesn't contribute to the glide characteristics of the model is thus jettisoned with the pop pod.

The pop pod shown in Figure 13-20 incorporates a streamer for recovery. Some of the larger pop pods for larger motors use parachutes. But I've learned from long and hard experience that a parachute will often cause the pod to get tangled with the glider. This happens often enough, anyway, even with a streamer recovery pop pod. It results in a "red baron." The whole model spins slowly to the ground with the streamer, pod, and glider tangled up, looking like the Red Baron himself shot the whole works out of the sky.

You want that pop pod to drop out of the glide path of the glider so that they don't run into each other and get tangled. A parachute tends to keep the pop pod up there in the glider's flight path.

Always use a motor hook to ensure the retention of the motor in the pod, because the basic action of a pop pod requires an energetic ejection of the streamer and nose. The reaction force of this forward-moving mass pushes the pop pod backward on the model, unlatching the simple balsa hook on the pylon that mates with the notch in the forward fuselage of the glider.

PAINTING AND FINISHING GLIDE-RECOVERY MODELS

Finishing a glide-recovery model is another area of controversy where there's very little solid data to point in one direction or another. Some modelers leave their gliders unpainted, sanding the glider smooth with no. 400 sandpaper. They maintain that the extra weight of a smooth finish coat detracts more from the glide performance than the aerodynamically rough finish. Others use glider dope, a mixture of 50 percent clear model airplane dope and 50 percent dope thinner, putting on several coats and sanding with no. 400 sandpaper between coats until a very smooth finish is obtained.

Still others use filler to remove all balsa grain, then paint the model with several coats of acrylic enamel with a spray gun or airbrush. The bright colors of these paint jobs often enable judges and timers to see the glider at a greater distance.

A popular finishing technique, especially in Eastern European countries and with advanced modelers flying with motors of Type D or larger, is to cover the balsa wing and stab surfaces with model airplane tissue, which is then doped to tighten and strengthen it. This can result in a very strong wing or stab.

The biggest problem faced by a modeler in finishing the relatively fragile parts of a glide-recovery model is the fact that model airplane dope shrinks as it dries. This can cause large, thin surfaces on wings and stabs to warp. Therefore, some modelers add a few drops of castor oil to a bottle of dope to destroy this shrinking characteristic. *But it doesn't work with all types of dope,* so run a few tests first.

TRIMMING

A glide-recovery model must be balanced or trimmed for glide before it's flown. There are very few glide-recovery model designs that can be flown without preflight glide trimming; delta-wing types often must be trimmed by means of a series of actual flights with low-powered motors.

Trimming a B/G is easy. Use the glider alone without the pop pod mounted or without the propulsion module installed—that is, the gliding portion must be trimmed in its gliding configuration. A grass-covered field is best for trimming because it keeps the glider from getting dinged if it dives into the ground. Naturally, if there's only one rock in the field, the glider will be drawn to it like a magnet and hit it. And, naturally, the glider will always break in the worst possible place.

The first trim flights should be gentle. Don't heave the glider! Grasp it behind the wings or in a place near the CG where you can get a good grip on it. Toss it gently away from you in a horizontal attitude with an overhand motion and release it into a glide path just *slightly* below the horizontal. Make several tosses to confirm what the glider does. You may not have tossed it correctly the first time.

If the glider dives as shown in Figure 13-21, it's nose-heavy. You must therefore remove some weight from the nose or add some to the tail, or both. If the glider stalls as shown in Figure 13-22, it's tail-heavy, and you'll have to add some weight to the nose. Tail heaviness is the most common pretrimmed condition of any glider. And that's why I always hollow out a compartment in the nose of my gliders so that I can put clay or solder weight inside the fuselage where it won't create a lot of drag.

The best thing to use for trim weight is a little glob of plasticene modeling clay, which can usually be purchased in the crafts section of the hobby store.

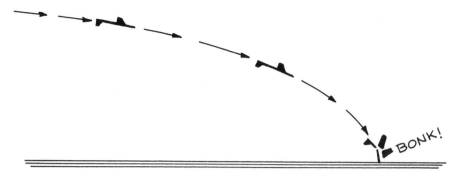

FIGURE 13-21 Nose-heavy gliding flight.

Put a hunk in your range box. To use it, pinch a little onto the nose or tail of the glider, as the case warrants. Or put it into the nose compartment as I do. Add a little bit of clay and make another trimming flight. Add or take away a little clay until the glider sails away from you in a slowly descending glide path. A normal B/G design should land 15 to 30 feet away from you, depending on how tall you are and on the design of the model.

Now add just a teeny pinch of clay to the *left* wing tip to cause the glider to turn slowly to the left as it glides. Once the glider gets a couple hundred feet in the air after boost phase, you don't want it to sail off on a straight cross-country mission. A straight-flying glider will nearly always turn its tail into the wind and take off for the next state, flying downwind much faster than you can possibly run. A motorcycle or dirt bike is very helpful in this situation. What you want is a glider that turns slowly over the launch area like a hawk circling its prey. Why a left turn?

On warm days, rising bubbles and columns of hot air, called thermals, are generated in most open fields because a hot spot on the ground heats the air directly above it. The heated air is less dense than the cooler air around it and it starts to rise. It may rise either as a bubble or as a column of air. A thermal is a small low-pressure cell like a miniature storm system, except it isn't big enough to have clouds in it. Thermal bubbles are like doughnuts,

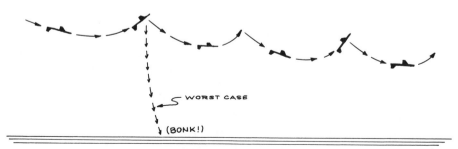

FIGURE 13-22 Stalling, tail-heavy gliding flight.

and thermal columns are weak dust devils and very weak tornado funnels. When viewed from below in the northern hemisphere of earth, they circle counterclockwise. A glider that turns right will stick its nose directly into the thermal airstream and stay in the thermal, rising with it. A left-turning glider will fly through or out of a thermal, and you'll be able to get it back. I've seen B/Gs lost during trim flights, disappearing upward after being caught in a thermal.

Once having trimmed for a gentle left-turning glide, the model's ready for a heave test. Haul off and heave the glider straight up as hard as you can. Good; the wings stayed on. The glider should climb straight up, losing speed, roll to horizontal attitude, and glide in a left circle about 25 to 50 feet in diameter. If it doesn't, try again. If it still won't perform right, remove just a *little bit* of nose weight and make it *slightly* tail-heavy.

Don't worry if your glider won't pass a heave test; some designs won't, no matter how hard you try. If you cannot get the blasted thing to roll out and glide after persistent attempts to get a successful heave test, go ahead and try it in a powered flight, anyway. Now balance the model for boost flight. Add the pop pod or propulsion module with an unused motor installed. The model should balance at or ahead of the leading edge of the wing or at the point the kit manufacturer specifies in the instructions.

(*You threw the instructions away after you built the model?* You didn't put the instructions in your notebook? Didn't you have the presence of mind to read the instructions thoroughly or to realize that you might need them when trimming the model? Didn't you think that you could build another model just like the first from parts if you traced the planforms on a sheet of paper and kept the instructions to tell you how to build it again?)

If a B/G model doesn't balance at or ahead of the leading edge of the wing or at the point specified by the manufacturer, add or remove weight to or from the pop pod on the nose of the pod. It's impossible to do this with a rear-motored model using a propulsion module, which is one reason rear-motored models aren't as popular with contest B/G flyers. If your rear-motored model is tail-heavy in boost configuration, you've got to take weight out of the propulsion module, not off the model, because the model has already been trimmed for glide.

If you've got a B/G or RG that uses surface change—that is, control surfaces that move at transition—check to make sure that the surfaces are locked into proper position for boost flight. If it's a kit, check the instructions concerning this.

Trimming an RG model is similar to trimming a B/G. Trim for glide first with an expended motor casing installed and everything in glide configuration. An RG may glide faster than a similar B/G because of the extra weight it carries, so you may not be able to glide-test it in any other way than tossing it from a second-story window to get enough altitude for an observable glide.

Carefully check the boost-phase balance of any RG. This is where most RG models get into trouble. The first launch of an RG should be a heads-up affair, because RGs have been known to loop and attempt to part the hair of the modeler or bystanders.

FLYING GLIDE-RECOVERY MODELS

Now you're ready to launch. Make sure the pop pod is free to come off at transition when the ejection of the nose and recovery package kicks it rearward. Make sure it won't come off during boost; dangle the model from the pop pod to check this.

In other types of B/G and RG models, make sure that the transition mechanism, whatever it is, is set to work as it's supposed to. Follow instructions for kit models.

A front-motored model normally needs to be held up off the base of the launch pad and jet deflector. A clothespin or piece of tape wrapped loosely around the rod (you may want to get the tape off for the flight of another model later) will support the model properly. Give it all the rod you can by making sure the tail of the model just barely clears the launch pad base or jet deflector.

Use an umbilical mast to hold the ignition wires and clips for a front-motored model as shown in Figure 13-23. An umbilical mast allows the leads

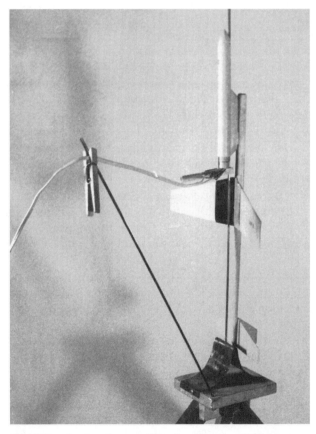

FIGURE 13-23 An umbilical mast installed on a launch pad to support the ignition wires and prevent them from fouling the tail of the B/G at liftoff.

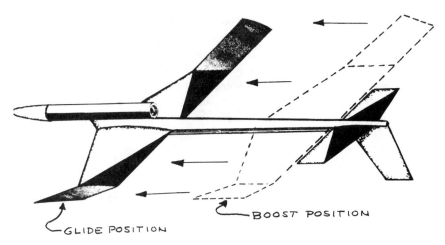

FIGURE 13-24 Kuhn's Buzzard movable-wing rocket glider (RG). The wing is held at the rear by thread during the boost phase. The motor ejection charge burns the thread and a rubber band pulls the wing forward on the fuselage into a position for glide phase. Nothing is ejected or dropped from an RG model.

to fall freely *away* from the model so that they don't get caught on the model's tail section, for example. You can also tape the leads to the launch rod to keep them from falling into the model, but be sure that the model will rise up the rod clear of the wires and tape. It's very embarrassing to have the leads tangle with the model, especially in a contest where the rules say that such an aborted launch counts as an official flight.

FIGURE 13-25 This radio-controlled glider was built by Bob Mosely. (Photo by Gary Rosenfield.)

If there's a wind blowing, it may push the model around on the launcher. Try to steady it with a clothespin or a piece of tape to prevent it from being blown or to keep the wind from blowing a B/G glider off its pop pod. But make sure the model will go up the rod upon ignition.

It's always a good idea to have somebody to help you on recovery, especially with a B/G. Your partner should go after the B/G pop pod or propulsion module while you chase the glider. Don't ever count on range people or bystanders to watch for your pod or booster. Basically, there are two types of glide-recovery model rocketeers: those with a large collection of pop pods with no gliders and those with a collection of gliders whose pop pods were never found.

Yes, it's possible to design, build, and fly glide-recovery model rockets using the time-honored model airplane technology of dethermalizers. A dethermalizer is an on-board timer device that does something to the model after a given number of minutes to destroy its good glide characteristics and make it sink more rapidly to the ground.

Now go out there and fly your B/G! You're involved in one of the most fascinating areas of model rocketry!

14

BUILDING AND FLYING LARGE MODELS

Although most model rockets use Type A through Type D motors, the "upper end" of the model rocket hobby embraces Type E through Type G motors. Some older and more experienced model rocketeers enjoy building and flying these large model rockets. However, larger models take longer to build and are more expensive. Type E through Type G model rocket motors are more expensive. Large model rockets require a bigger flying field.

And because of the increased amount of rocket propellant in the larger motors, the higher weights of the models, the higher velocities attained, and the greater altitudes possible, *all* the safety rules must be followed *all* the time. I have emphasized safety strongly throughout this book. I want the NAR Model Rocket Safety Code to become second nature to you when building and flying all types of model rockets. Even when flying the smallest models, you should follow the safety code because of the good habits this forms. If you fly larger model rockets, you will need every bit of safety consciousness and safety training that you've acquired while flying the smaller ones.

Everything that applies to smaller model rockets also applies to large model rockets—and usually more so. You may wish to go back and review some of the earlier chapters as you get into this one. Don't be embarrassed about doing it. No one but you will know you're rereading the earlier material. Better to do this now than to have a disaster with a $50 model traveling at the speed of sound.

CONSTRUCTION

Large model rockets generally use many of the same construction techniques and materials as the smaller ones. However, special care must be taken to make the model stronger and more durable. A large model rocket propelled by a model rocket motor with more than 20 newton-seconds of total impulse will undergo high accelerations. It will achieve higher airspeeds,

FIGURE 14-1   A growing selection and variety of large model rockets are available for the advanced model rocketeer. (Photo by AeroTech Consumer Aerospace.)

perhaps approaching the speed of sound (Mach 1). Therefore, it will experience some high aerodynamic forces because, if you recall, drag and lift forces increase as the square of the airspeed. Some of these large model rockets will attain speeds 4 to 5 times higher than their smaller brothers; therefore, the forces on their fins, for example, will be 16 to 25 times as much!

Large model rockets must be built *strongly*. In most cases, stronger construction techniques must be used. Stronger and heavier materials are required. And, of course, good glue joints and careful construction are mandatory.

Large noses up to 6 inches in diameter are available from many model rocket manufacturers. Some of these are large versions of the small blow-molded polystyrene noses found on small kit models. Others are turned wood. They are expensive. A 3-caliber ogive with a 4-inch base diameter can cost more than $15. If you intend to build and fly a lot of large model rockets, you might consider investing in a wood lathe. It may pay for itself after making five to eight large noses.

Large-diameter spiral-wound paper body tubes are available with diameters up to 6 inches. However, most large model rocket kits use tubes with diameters of 1.9 inches and 2.6 inches. Both AeroTech Consumer Aerospace and Estes Industries, Inc., sell large body tubes along with matching noses of various shapes and internal parts such as motor mounts and bulkheads that fit their tubes.

You may be tempted to use one of the heavy paper tubes on which carpets and linoleum are wound. My suggestion is *don't*. They are heavy, and they aren't very strong. I've seen them collapse under the thrust of an F50 motor, so they certainly won't take the kick of a G80!

In large-diameter body tubes, it is important to use a stuffer tube. If you recall, this is a tube of smaller diameter just large enough for the motor being used. It is centered in the larger tube from the tail up to where the parachute compartment begins. It is held in the large tube by model airplane plywood, fiberglass, or heavy cardboard centering rings. The stuffer tube reduces the internal volume that must be pressurized by the motor's ejection charge. Often, the body tube on a large model rocket is so big that the puff of gas from the ejection charge can't build up enough pressure to pop the nose and deploy the chute. The stuffer tube ducts all the ejection charge gases into the parachute storage compartment. The large, empty volume of the body tube doesn't have to be pressurized. In a sense, you're flying a lot of dead air in there.

The motor mount must also seal the ejection charge gas to prevent it from escaping to the rear, thus also causing a failure to deploy the chute. Motor mounts for Type E, Type F, and Type G motors must be *robust*. A Type G80, for example, produces a peak thrust of 24 pounds, and this is a *lot* more than the 2 pounds of the 18-millimeter Type A through Type C motors of the smaller models. Ejection charges in the larger motors are also more energetic because they have to push a heavier silk or nylon parachute out of the tube and perhaps push off a heavy nose cone carrying a payload. You should use a motor retaining clip that must be larger and heavier than the ones used on smaller models. Otherwise, the motor may be kicked out instead of the recovery system. So the motor mounts and retention devices in a large model rocket must be extra strong.

Large model rockets often have heavy nose and payload sections. When such noses are popped off in flight, they come up against the end of the shock cord with a considerable jerk. Therefore, heavy-duty shock cords must be used, and they should be made longer than those in small model rockets. Visit a fabric or notions store and get wide elastic that is used to make waistbands in clothing, for example.

Shock cord attachments must also be robust. A simple screw eye in the base of a balsa nose will usually pull out. So inset a ½-inch wooden dowel into the nose base, glue it in, and screw a large screw eye into it. Shock cords should be affixed to the main body of the model by gluing them through one of the bulkheads such as the one at the base of the parachute compartment. Many designers attach a piece of stranded stainless steel wire to the motor mount and affix the front end to the elastic shock cord; this prevents the ejection charge from burning the elastic shock cord in two. A separation with a 1-pound model isn't funny because either the nose section or the main body

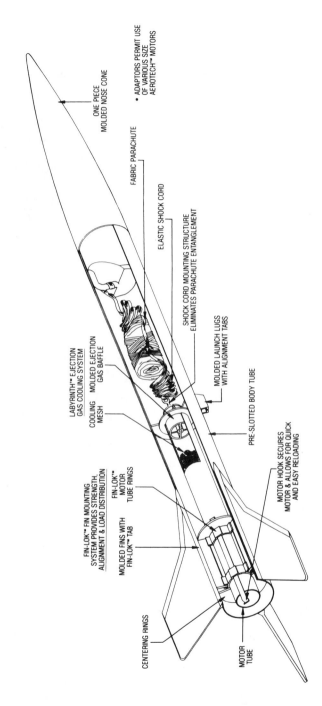

FIGURE 14-2 Cutaway drawing of the AeroTech Initiator showing the typical heavy-duty construction of large model rockets. (Courtesy AeroTech Consumer Aerospace.)

ONE PIECE MOLDED NOSE CONE

• ADAPTORS PERMIT USE OF VARIOUS SIZE AEROTECH™ MOTORS

FABRIC PARACHUTE

ELASTIC SHOCK CORD

SHOCK CORD MOUNTING STRUCTURE ELIMINATES PARACHUTE ENTANGLEMENT

MOLDED LAUNCH LUGS WITH ALIGNMENT TABS

PRE-SLOTTED BODY TUBE

LABYRINTH™ EJECTION GAS COOLING SYSTEM

MOLDED EJECTION GAS BAFFLE

COOLING MESH

FIN-LOK™ FIN MOUNTING SYSTEM PROVIDES STRENGTH, ALIGNMENT & LOAD DISTRIBUTION

FIN-LOK™ MOTOR TUBE RINGS

MOLDED FINS WITH FIN-LOK™ TAB

CENTERING RINGS

MOTOR HOOK SECURES MOTOR & ALLOWS FOR QUICK AND EASY RELOADING

MOTOR TUBE

may not tumble as it descends. A DSE (detectable seismic event) with a large model rocket is a fearful thing to witness.

Although I have built some large model rockets with sheet balsa fins, I don't recommend this material as fins for large model rockets. Sheet balsa can be used as the *core* of a fin if a doped tissue or Monokote covering is shrunken over it to provide additional strength. Fins should be built using model aircraft plywood at least ⅛ inch thick and preferably ¼ inch thick. Some modelers have used fiberglass sheet.

Put a good airfoil on the fins of large model rockets. This means rounding the leading edge and tapering the trailing edge as we discussed earlier. This will reduce the drag force on the fin and make it more effective. It will also decrease its weight.

AeroTech Consumer Aerospace has seven different injection-molded plastic fins that reduce design and assembly time. These fins can be customized by cutting them with a *hot knife*. This is a small soldering iron to which an X-Acto knife blade has been attached at its tip. The hot knife slices plastic like butter (usually) so that you can cut your own shape. Or you can cut your own fins out of sheet styrene that is available in most hobby shops. A file can be used to shape the fin, and sandpaper will smooth it. After you paint it, no one will know that it's a custom-cut fin. I've built several large scale models using sheet styrene, a hot knife, and a file to make my own heavy-duty fins.

Fins should not be attached to large model rockets using the common butt joint of smaller models where the fin root is simply glued to the external surface of the tube. Through-the-wall fin attachment must be used. This involves cutting a slit in the body tube and designing the fin with a tab or tenon that extends through the body tube to be glued to the motor mount or stuffer tube, for example. AeroTech Consumer Aerospace uses a patented FIN-LOK ring on the motor mount/stuffer tube that holds the fins in place. You can build a similar system where the fin tab fits between a forward and an aft bulkhead and is centered by two strips of basswood glued along the motor mount/stuffer tube. Always use a fin fillet because this adds to the strength of the fin as well as reducing interference drag.

Launch lugs should be securely attached to the body tube. To add strength, I often dope a sheet of tissue over the lug and the adjoining area where it is attached to the body tube. I've also reinforced fin roots using model airplane tissue doped in place. See Figure 14-4.

What sort of glue works best on these large model rockets? If you make a good glue joint, almost any glue will work. However, most large model builders use epoxy, which is a two-part bonding agent. If you use epoxy, use the stuff that cures in an hour or more. The fast-curing 5-minute epoxies aren't as strong as the ones that take longer to cure. And you want that glue joint to be *strong*!

Because an unstable model rocket powered by a Type G motor is downright terrifying and even potentially deadly, check and recheck your CP-CG calculations. If in doubt, give the model the swing test. Make certain that it is stable.

FIGURE 14-3 Through-the-wall fin attachment. The tab on the fin goes through the slot in the body tube and is bonded to the inner motor tube to provide added strength to withstand the forces created by higher airspeeds.

Whatever you do for smaller model rockets, do it double for the big ones. Make sure all glue joints are good. Make sure the CG and CP are in the proper relationship. Make sure the recovery device is installed with strong attachments. Don't take shortcuts or do sloppy work on large model rockets. They can bite.

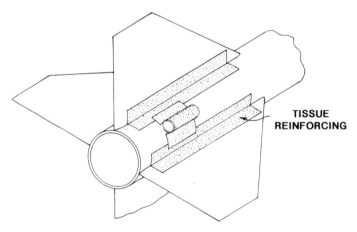

TISSUE REINFORCING

FIGURE 14-4 Reinforcing the fin roots and launch lugs with tissue.

IGNITION OF CLUSTERED MOTORS

Spaceships and expendable launch vehicles often require more thrust than is available from a single-rocket motor. Therefore, engineers have designed vehicles with clusters of rocket motors that are ignited simultaneously. The NASA space shuttle uses a complex cluster of three liquid propellant engines and two solid rocket boosters. The single-stage-to-orbit spaceships also use clusters.

In the early days of model rocketry, we developed clustered motor technology because we had only Type A and Type B motors. And for large model rockets we needed more thrust than a single motor could provide. The big problem with all clustered rocket motors is achieving near-simultaneous ignition of all the motors. The model should leave the launch rod with all motors ignited and thrusting. This was and still is no easy feat to accomplish.

From 1970 to 1990, the use of clustered motors by model rocketeers dropped to nearly zero. Few kits were available. Cluster ignition was tricky, complex, and unreliable. We found it easier, simpler, more reliable, and safer to use a single larger motor in place of a cluster of small ones. Now clusters are back for large model rockets. Because cluster ignition technology is difficult and advanced, I've put it into this chapter on building and flying large model rockets.

The most common cluster is three model rocket motors grouped together in the tail of a large body tube as shown in Figure 14-5. Three 18-millimeter motors will neatly fit into a tube 1.63 inches in diameter.

A bit of model rocket history: Anyone care to guess where we got the body tubes for the first three-motor cluster rockets? Answer: the cardboard center core tube from a roll of paper towels. Three 18-millimeter motors would fit neatly inside. Later, this tube diameter became the Estes BT-60. And it soldiers on today as various tube designations from various manufacturers.

In a cluster, each motor is mounted in its own motor mount tube with a thrust ring and a motor clip. The space between the motor tubes and the main body tube is closed off with a bulkhead to prevent the ejection charge gases from escaping to the rear.

Ignition of the motors in the cluster is tricky, and your ordinary electrical firing system *will not do it!* You're going to have to build one that will use a 12-volt motorcycle battery. Or you can use your existing system if you build a relay ignition system as an add-on for clustering.

The only workable way to wire up the igniters for clustered motors is in *parallel*, not in *series*. These are electrical terms that describe the wiring in a circuit. *Series* hookup is igniter to igniter to igniter, so electrical current flows through each of them in series. The big problem here is that once one igniter activates and breaks the circuit, electrical current stops flowing through the other igniters. That you do not want. You want the current to continue to flow through all the igniters until they are all activated.

This requires a *parallel* hookup as shown in Figure 14-5. Make a special clip whip as shown. Different-color wire—red and black, for example—should be used for the two whips. Then when you're under that cluster model

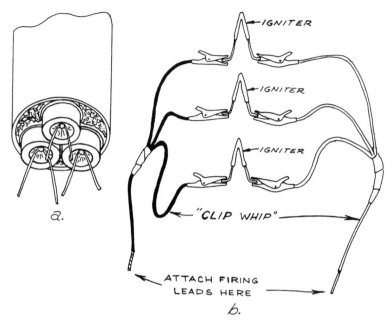

FIGURE 14-5 (a) A cluster of three model rocket motors with their igniters installed; (b) parallel hookup of the igniters using clip whips.

trying to hook up all the wires in that rat's nest of wire, you know to hook one red wire and one black wire to each motor igniter. Make sure the clips don't touch one another and short out.

When doing a cluster, I use a relay add-on, whose wiring diagram is shown in Figure 14-6. Each hooked-up clip whip is attached to one of the two leads coming from the relay contacts going to the rocket as shown. The ignition leads from your regular firing system are hooked across the relay coil as shown.

Finally, just before you leave the launch pad, wires are connected from the relay contacts to a large battery right at the launch pad itself. This battery should be a high-capacity one because it will have to fire three or more igniters at once, and this takes three or more times the electrical current that it takes to fire a single igniter.

The relay system is activated by your own regular firing system. When you push the firing button, the current from your regular system actuates the relay coil. This switches the relay contacts and puts the current from the big battery right into the igniters. Because the large battery is next to the launch pad, the wire lengths are very short and there is very little voltage drop due to the wiring.

The relay should have a 6-volt DC coil if you're using a 6-volt firing system run with four AA penlight cells. If you have a 12-volt firing system, use a 12-volt relay. Actually, a 6-volt relay can be used for both because 12 volts

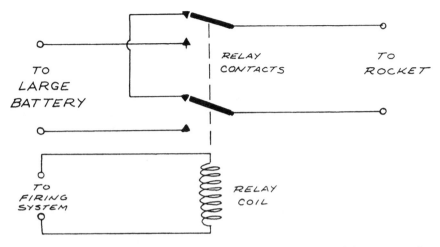

FIGURE 14-6   Schematic wiring diagram of the relay-type firing system for ignition of clustered-motor model rockets.

will activate a 6-volt relay. Always use a relay with a DC coil because a relay with an AC coil will not work with the direct current of an electrical firing system. The firing system should be wired to the coil. The normally closed contacts should be shorted together. The normally open contacts should go to the launch battery. And the center pole contacts should go to the rocket and clip whips.

The relay should be housed in a small aluminum or plastic box with wire leading to terminals on the box. The box should be sealed to keep dirt away from the relay contacts. Radio Shack also has suitable aluminum or plastic boxes that will hold the relay and the terminals. Be sure to mark which terminals are which so that you don't get confused and hook the wrong wire to the wrong relay parts.

The relay can be thought of as a remotely operated firing switch. The passage of electrical current from your launch controller goes through the relay coil and activates this remote switch. A relay system doesn't cost very much, it can be used over and over again, and it helps ensure reliable cluster ignition.

Again, I must stress that care and patience are required in both the construction and the flight of clustered motor models. If one motor in a cluster does not ignite, the thrust will be low and unbalanced, and the model will veer in flight. Because of the possibility of ignition failure with clustered motors, motors should *never* be placed out at the tip of fins; if one motor fails to ignite, the off-balance thrust will cause the model to pinwheel in flight once it leaves the launch rod.

How many motors can be successfully ignited in a cluster? I've done seven, but I know of other model rocketeers who have ignited very large clusters. But beyond a cluster of three, the additional weight plus the very large base of the model tend to work against you; you're better off using a single larger motor.

FLIGHT OPERATIONS

Although I have launched model rockets powered by Type F motors from $1/8$-inch rods 36 inches long, I prefer to use at least a $3/16$-inch rod 60 inches long. Some manufacturers require a $1/4$-inch rod for launching their large and heavy kit models.

The NAR Model Rocket Safety Code requires that you have at least 30 feet between you and the launch pad. You should stay at least 30 feet back and require that all spectators do the same. If something goes wrong with that large model out there on the pad, distance is safety.

Follow the ignition instructions of the motor manufacturer. Some Type F and Type G motors require special igniters and ignition techniques.

Carry out your preflight prep with extra care. Have someone double-check your work. With large model rockets, you have no margin for error.

You will need a large flying area. A school football field isn't large enough for any model propelled by more than 20 newton-seconds (Type E motor or larger). Get out of town and into the country. The NAR Model Rocket Safety Code requires that you have a launch field with a minimum dimension of 1,000 feet for Type E through Type G motors. You can stage and cluster motors up to 320 newton-seconds if your field has a minimum dimension of at least 1,500 feet.

Tilt the launch rod a few degrees away from the vertical and toward an open and unpeopled downrange area where the model can prang if something goes wrong. The downrange recovery area should be clear of people, and you should keep it clear during large model rocket flight operations.

Although smaller model rockets can be placed on launch pads a few feet apart, put your large model rocket launch pad about 10 feet from other launch pads. When ready to launch, make sure everyone in the area is aware of the pending flight. No one should be lying on the ground. Motor storage boxes should be shut. Everyone's attention should be on the big model.

The liftoff will be spectacular! The sound and fury will be unbelievable! The model will probably go out of sight overhead. This is the time you should check for any possibility of a grass fire around the launch pad.

Most large model rockets will go more than 1,000 feet and can top 3,000 feet easily. There's not a great deal of sense in trying for higher altitudes. Once you lose sight of the model, it's probably lost unless your tracking crews are using binoculars to follow it.

It may seem that I'm afraid of large model rockets. I'm not afraid of them, but I do have a great deal of respect for them. A model powered by a Type D motor isn't as hairy as one powered by a Type G. But keep in mind that a Type G motor has roughly 8 times the power of a Type D and 16 times that of a Type C.

When flying large model rockets, you must be aware of the regulations of the Federal Aviation Administration (FAA). Model rockets weighing less than

1,500 grams (53 ounces) and propelled by less than 125 grams (4.4 ounces) of propellant are exempt from FAA control (FAR Part 101, Section 101.22). If your rocket weighs more than 1,500 grams, you must contact the FAA and file a waiver for these limits (FAR Part 101, Section 101.25).

Model rocketeers have accepted these voluntary limits on the size and weight of models and have always cooperated with public safety officials. There's more to it than that, however. There is financial and technical sense to these limitations. A high-performance Type G motor can cost up to $20 per flight, and a large model rocket kit can cost more than $50. When you start talking about big, high-powered, high-altitude rockets, you're talking about high-powered costs, too.

However, altitude performance isn't the basic reason large model rockets are built and flown. Large model rockets are basically payload carriers. They can lift miniaturized payloads to several thousand feet. And these payloads themselves can do fascinating things.

15

PAYLOADS

Most model rockets are built strictly for sport flying or competition. They carry no payloads other than their own airframes. But one of the reasons for the existence of present-day full-size rocket-powered vehicles is their ability to lift payloads to very high altitudes in very little time. Dr. Robert H. Goddard's original 1919 treatise on rockets written for the Smithsonian Institution was entitled "A Method of Reaching Extreme Altitudes." Goddard saw the rocket first as a means of carrying scientific instruments to altitudes that couldn't be reached by airplanes or balloons, and only second as a means of space travel. In times past and even today, rocket vehicles carry payloads such as explosive warheads, signaling devices such as flares, rescue components such as ropes and lines, and scientific instruments. The most important payload carried by rocket vehicles is people.

In model rocketry, we don't work with explosive warheads. This is forbidden by the NAR Model Rocket Safety Code and by all the rules, regulations, laws, statutes, and ordinances relating to nonprofessional rocket activity. Explosive warheads are very dangerous, and direct handling of explosives is not part of model rocketry. A great deal of highly specialized training is required in order to handle explosives with any degree of safety. People must be completely familiar with fusing and arming procedures—something they learn in military service after many years of training and experience. Explosives are far too hazardous for the average person to handle. If you're fascinated by explosives and things that go bang, you're reading the wrong book. Join the armed services and get the proper training as an explosives and demolition expert. Don't use model rocketry as your training ground. In model rocketry, we fly for fun and knowledge, not to conduct a small war.

Payload-carrying model rockets are the province of the experienced, advanced model rocketeer, not the beginner. Many things have been done with payload-carrying model rockets, including pollution patrol, smog control studies, and investigations relating to ecological factors, because model rockets are

FIGURE 15-1   A model rocket can lift various kinds of
payloads such as a fresh hen's egg.

a fast and inexpensive way to lift small cameras or temperature sensors to several thousand feet. Many things can be done with payload-carrying model rockets that haven't been done yet. So it's a wide-open field for careful research, creative development, and extensive testing.

Model rockets aren't usually designed to carry any old kind of payload that happens to come along, although they can and often are adapted to carry a wide variety of different payloads. Usually, a model rocket is designed around its intended payload, with the designer keeping in mind the size, shape, and weight of the payload plus any environmental factors that will affect the payload—acceleration, shock, vibration, heat, etc.

In nearly all cases, a model rocket's payload is carried in the nose section of the model. It may be housed inside a hollow nose or placed behind the nose in a cylindrical compartment that's a structural part of the nose assembly that comes off the model at ejection.

By positioning the payload in the nose, a better CG-CP relationship can be obtained due to the payload weight. This can be important in many payload models because the additional payload weight usually causes the model to have a lower liftoff acceleration and velocity, both of which require the model to have excellent stability characteristics.

The different payload types that are commonly carried in model rockets can be grouped into the following general categories:

1. Passive payloads
2. Optical payloads such as cameras
3. Electronic payloads
4. Biological payloads
5. Special payloads

Although some of these categories overlap, we can discuss them individually because their airframe and propulsion requirements are often quite different. You'll see for yourself where the overlaps exist.

PASSIVE PAYLOADS

When model rocketry began in 1957, the main emphasis was on propulsion and airframe technologies because *everything* was new. We had to begin with very simple model rockets and very simple technologies. We didn't have a large selection of model rocket motor types from which to choose. In late 1957 and early 1958, we had only Type A4 motors, and we got Type B4 motors late in 1958. Therefore, we didn't have the propulsion capability to lift payloads. Small payloads were a real challenge because transistors were just becoming commercially available and integrated circuit chips hadn't even been heard of yet. Today's miniature digital cameras were also years in the future.

The NAR recognized the basic payload-carrying potential of model rockets and developed what was then called the "passive payload" competition category, now called the payload category. It's based on the ability of a model rocket to carry one or more standard payloads to as high an altitude as possible with a given maximum amount of total impulse.

The early standard payload was a small wafer of lead with a diameter of 19.05 millimeters (0.75 inch) weighing no less than 28.35 grams (1.0 ounce). It was selected in 1959 because the smallest available body tube had an internal diameter of 0.75 inch. Later, the dimension and weight were converted into the metric system for international competition.

In 1979, the NAR Contest and Records Committee changed the standard payload because they believed that the lead payload could be dangerous (although no accidents had ever occurred with a model rocket carrying this payload). The current NAR standard payload is a nonmetallic cylinder containing fine sand with a weight no less than 21.0 grams and a diameter of 19.0 millimeters, plus or minus 1.0 millimeter.

The contest rules require that no holes be drilled in the payload, that it not be altered in any way, that nothing be permanently attached to it, that the contestant be able to insert and remove it from the model at will, that the payload be totally enclosed within the model's airframe, and that the model be so designed that the payload cannot separate from the model in flight.

In essence, the rocketeer is being put in the shoes of a real rocket design engineer who's told, "Here's the payload. It weighs this much and has these dimensions. You cannot change its shape. You cannot alter it in any way. You don't even have to know what it is or what it does. Just build a rocket vehicle to take it as high as possible and get it back unharmed."

These specifications aren't very difficult to meet. Nearly any good sporting model can be converted to a passive payload model by adding a payload compartment. But it probably won't win any contests. The model rocket designer tries to achieve the optimum design for a model that will meet all the requirements of the competition rules.

A typical payload contest model is shown in Figure 15-2. In view of what we've learned in the preceding chapters about aerodynamics and shapes, note

FIGURE 15-2   Two different contest models for the NAR Payload event. The standard payload is the black tube on the lower right.

the little extra touches that contribute to the increased overall performance by means of drag-reduction techniques. The model has been optimized for one purpose only: to carry the standard payload to the maximum possible altitude. Aerospace Specialty Products (see Appendix I for the address) sells a competition payload model kit and the standard payload module.

I've built and flown a lot of competition payload models over the years. It's impossible for me to say, "This particular design is the very best payload model design possible." Too many trade-offs have to be made, and every rocketeer makes them differently. I *do* know that a short, squatty model will oscillate too much in flight, as we've previously discussed. I also know that a payload model must have very efficient fins that will produce a strong fin restoring force, such as the shape shown in Figure 15-2. I also cannot tell you whether the fin shape shown in Figure 15-2, the long-span clipped delta, is better than a clipped delta with 60 degrees of leading edge sweep. This latter fin has less normal force but stalls at a higher angle of attack, whereas the long-span clipped delta has a high fin normal force but stalls at a lower angle of attack. Why the emphasis on angle of attack? Because the payload model with its heavy nose has a lot more angular momentum than an ordinary model rocket. Therefore, it has the opportunity to swing to a much higher angle of attack before the fins can produce enough restoring force. The competition payload model has a higher moment of inertia—that is, once it starts to rotate in pitch-yaw, it develops considerably more rotational momentum than a light sporting model.

Payload contest models are fun. Putting 21 grams of payload up in the nose of a small model rocket *really* changes the way the model behaves. This is why it has been a competition event ever since the beginning of model rocket competition.

OPTICAL PAYLOADS

One of the most interesting model rocket payloads is a camera. Many model rocketeers have worked very hard to build and fly camera-carrying models.

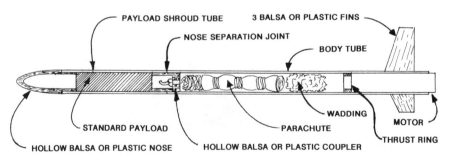

FIGURE 15-3   Cross-section drawing of a typical high-performance competition model rocket built to carry the standard payload.

The first camera-carrying model rocket on record was built and flown by Lewis Dewart of Sunbury, Pennsylvania, in 1961. A small Japanese camera was simply strapped to the side of the model rocket. When the ejection charge popped the nose, it also pulled a string that released the shutter, permitting the camera to take a picture of the ground below—or the sky and clouds, depending on the direction in which the model was pointed at the instant of ejection.

Estes Industries, Inc., brought out the first commercial model rocket camera, the original Cameroc, in 1965. The Cameroc lens pointed straight up through a transparent window in a hemispherical nose. Therefore, the model had to go over peak altitude and be pointed nose-down when the motor ejection charge popped the Cameroc nose off, releasing the shutter. The Cameroc took a single black-and-white photo 1.5 inches in diameter. In 1979, Estes introduced a greatly improved camera rocket, the Astrocam 110, which uses Kodak Kodacolor 110 color print film. The photos can be developed exactly the same as other Kodak 110 color print film. An improved Astrocam came out in 1992 with a preassembled camera and an improved lens.

As of the date of this updated edition, there are no commercially manufactured digital camera payloads available. However, there is a growing number of small, lightweight, and inexpensive general-use digital cameras on the

FIGURE 15-4   The Estes AstroCam 110 camera rocket takes a single color photo on each flight using Kodacolor 110 film. (Estes Industries, Inc., photo.)

market, and many rocketeers have taken to converting them to useful camera payloads. The best I've seen was engineered by NAR member Bob Kaplow. He converted a simple fixed-focus, fixed-exposure Aiptec Pencam that shoots twenty-six VGA pictures stored in a built-in 8 MB RAM. After flight, the pictures are downloaded via a USB cable.

It has also occurred to model rocketeers that a motion picture camera in a model rocket would produce a spectacular piece of footage as the ground fell away and the model climbed into the sky. The first movie camera was flown in a model rocket by Paul C. Hans and Don Scott of Port Washington, New York, in 1962. It was a big model for its time, powered by the most powerful model rocket motor then available, the Coaster Type F blackpowder unit. The smallest motion picture camera available at the time was the Bolsey B-8, a spring-wound 8-mm camera. Following months of design and testing, including flight tests carrying a dummy camera, Hans and Scott committed to flight.

Everything worked perfectly. Scott had to climb a tree to get the camera back. The color film was sent to the processing lab—and was lost! The company replaced the film but couldn't replace the flight footage. Hans and Scott tried it again at the Fourth National Model Rocket Championships (NARAM-4) at the U.S. Air Force Academy in Colorado Springs. This time, Hans took the film

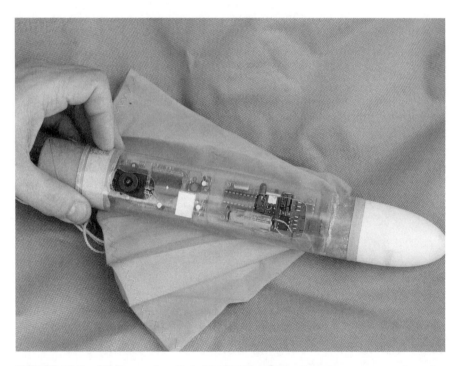

FIGURE 15-5   NAR member Bob Kaplow configured a clever conversion of a small digital LCD Pencam. Pictures are downloaded after flight via a USB cable. (Photo by Ric Gaff.)

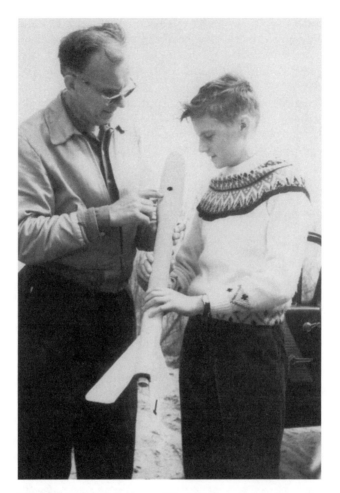

FIGURE 15-6   Charles and Paul Hans prepare the original movie camera model rocket for its first flight.

personally to a different processing lab with very explicit instructions. That first
model rocket in-flight motion picture film is indeed spectacular to watch. Paul
Hans still has it.

In 1969, Estes introduced the Cineroc, a small, lightweight Super-8 model
rocket motion picture camera developed by Mike Dorffler. The Cineroc was one
of the most elegant model rocket products ever put on the market. Thousands
of Cinerocs were flown by model rocketeers all over the world. The Cineroc was
available for about ten years, then Estes stopped making it. An Estes Cineroc
today is such a prized item that one will sell at an NAR auction for more than
$200. (I have two original Cinerocs carefully put away for posterity.)

Since then, model rocketeers have gone back to designing their own model
rocket movie cameras. However, the advent of larger model rockets and more

powerful model rocket motors have led some rocketeers to design and build on-board television cameras that send the video signal back to the ground where it can be viewed in real time and recorded on a VCR. Chuck Mund was the first to do this, and it remains a very advanced area of payload model rocketry today.

With increasing electronic miniaturization and tiny CCD video cameras, it is now possible to fly an on-board television camera system. Although none are available commercially, you'll often see model rocketeers with their own conversions of tiny surveillance camera systems in their rockets at the flying field.

ELECTRONIC PAYLOADS

The possibility of launching a television camera leads directly to the subject of electronic payloads. Real research rockets carry small radio transmitters that receive the electrical output of sensors—temperature and pressure pickups, for example—and send the data back to the ground by radio link. This is called *telemetry*. The Germans began to develop it at Peenemünde during World War II when they were flight-testing the V-2 rockets. They didn't have telemetry at first. They quickly developed a system that would radio back to the ground the data from on-board instruments. One day after the third production V-2 blew up 3 seconds after ignition, the chief test engineer muttered to another test engineer, "We just blew a million deutsche marks in order to guess what could have been reported accurately by something worth the price of a small motorcycle!"

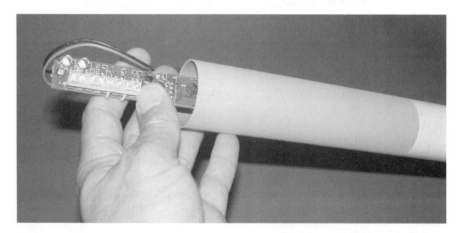

FIGURE 15-7 The G-Wiz LC standard flight computer uses proprietary firmware algorithms to determine the key events in a rocket's trajectory such as launch, booster burnout, sustainer burnout, coast, apogee, and low altitude.

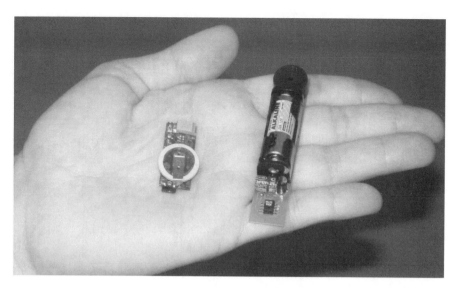

FIGURE 15-8   Two of the smallest altimeters available. On the left is the Alti-
mate 1 from BonNova. It weighs 4.5 grams with battery and can record altitudes up to 30,000 feet AGL. On the right is the Adept Instruments Model A1. It
records altitudes up to 3,000 feet AGL.

Model rocketeers don't blow a million dollars when they launch an experimental model rocket, but many of them would like to know the speed or
acceleration of the model. Or the temperature of the surrounding air on the
way up.

The first model rocket transmitter was designed and built by John S. Roe
and Bill Robson in Colorado Springs in August 1960. About a year of work
went into its design because transistors had just become available. The initial

FIGURE 15-9   AED Electronics R-DAS mounted on its payload bay carrier.
(Photo by Ken Horst.)

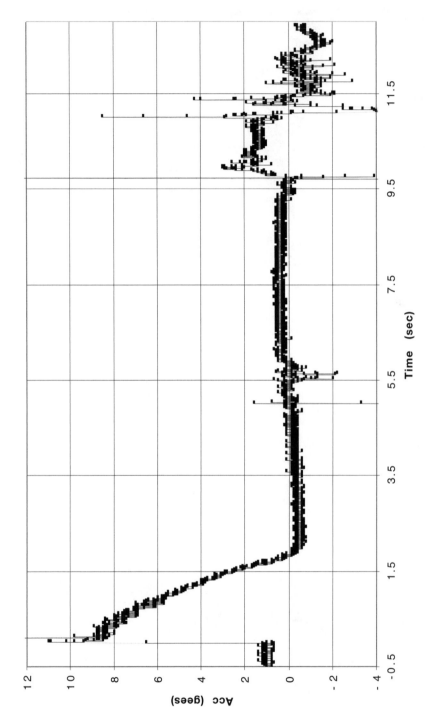

FIGURE 15-10 R-DAS acceleration data. Tomahawk on a G40-10. (Excel chart by Ken Horst.)

version was a simple oscillator broadcasting in the 27-megahertz Citizen's Band. It merely said, "Here I am!" By using a directional antenna, the model could be located on the ground after it landed. Roe and Robson also developed a transmitter that could be hooked to a photovoltaic cell and thus report spin rate every time the photocell looked toward the sun.

Today, there are a number of useful electronic payloads available to rocketeers. Actual on-board computers known as flight computers can control everything from upper-stage ignition of composite motors to parachute deployment at apogee or even parachute deployment at low altitude (this greatly reduces the distance you have to run/walk to recover your rocket). Besides using an on-board computer to control various functions of your rocket during flight, computers can also be used to accumulate data. The most common of these are simple barometric altimeters. Adept Instruments makes one of the most reliable and easy-to-use models. The A1 measures altitudes to 3,000 feet above ground level (AGL) and costs less than $60. After a flight, an A1 altimeter will be beeping out the maximum altitude in this manner: (1,325 feet) Beep . . . Beep Beep Beep . . . Beep Beep . . . Beep Beep Beep Beep Beep. See Figure 15-8. John Coker hosts a useful Web site that compares features of most of the currently available altimeters at www.jcrocket.com/altimeters.shtml.

AED Electronics manufactures an on-board electronic R-DAS (rocket data acquisition system). Take a look at Figure 15-10. It shows the acceleration data from an Aerotech Tomahawk flown with a G40-10. The R-DAS was programmed to record data 200 times per second (200Hz).

Because modern solid-state integrated-circuit electronics is becoming more sophisticated and cheaper daily, we can look forward to seeing more and more on-board electronic payloads in the near future. The sky is going to be the limit, and we're in for a lot of fun as these on-board electronic modules become more ubiquitous. I can hardly wait. . . .

BIOLOGICAL PAYLOADS

Sooner or later, a model rocketeer gets the bright idea that it would be fun to fly a mouse in a model rocket. This is known as live biological payload, or LBP for short. Don't do it!

A lot of mice have been killed in model rockets, proving only that a model rocket is a very expensive mousetrap. But it doesn't work as well as an ordinary mousetrap because you must first find a mouse and entice the little beastie into the model rocket's payload compartment. Mice bite. And they fight, too.

Unless you are conducting a valid scientific experiment under the direct supervision of a biology teacher, there is *nothing* you can learn by flying an LBP. You can subject a biological specimen to the same environmental stress on the ground without a rocket flight at all.

The NAR does not support the flying of live animals in model rockets, and an NAR member can lose membership privileges for doing it. The

American Society for the Prevention of Cruelty to Animals (ASPCA) has even stronger objections to model rocketeers flying live animals. The ASPCA has gotten *court orders* that completely shut down *all* model rocket activity in several schools, clubs, and even towns because a mouse flight was reported in the news media.

Don't fly an LBP. There's a better way that's more fun.

SPECIAL PAYLOADS

I couldn't think of a more technical term than this to describe something called "egg lofting." In 1962, Captain David Barr of the U.S. Air Force Academy proposed an idea that was immediately adopted by model rocketeers. Captain Barr felt that an excellent test of a model rocketeer's ability would be the flight and recovery of a *fresh* grade A large hen's egg without cracking the shell.

A fresh grade A large hen's egg weighs an average of 2.7 ounces (76.5 grams) and has a minimum diameter of 1.75 inches (44.5 millimeters). Today's quality-controlled egg factories produce these eggs with close-tolerance weights and dimensions. However, these mass-production eggs also have very thin shells, and this adds to the challenge of egg lofting. Not only is an egg a very fragile payload but if something goes wrong you have a scrambled egg and not a dead animal.

The first successful egg flight was made by Paul Hans and Don Scott in their movie camera rocket at NARAM-4 in 1962. Since then, tens of thousands of eggs have taken to the air in model rocket flight, and we've learned a lot of tricks about doing it.

An egg is very strong in its long dimension and very weak in its minimum diameter. However, if it is solidly cushioned all around, it will withstand a terrific beating. Essentially, you must put a strong shell all around the weak egg shell, an external structure that won't deform to crack the egg and that will *completely* cushion the egg all around to spread the impact stress.

Today, egg capsules, often called "hen grenades," are available as kits. The Quest "Courier" and Aerospace Specialty "Eggstravaganza" are both designed to hold and fly a single grade a large hen's egg. Special molded capsules are also available from Pratt Hobbies if you wish to design your own booster.

The egg capsules on these models are gems of careful design. I have seen such an egg capsule fall freely more than 1,000 feet out of the sky after the parachute stripped—and it landed and bounced. The egg was not cracked.

Although such egg capsules have taken some of the fun out of egg lofting, someone always does it wrong. Someone regularly prangs an egg capsule in such a way that white and yellow goo goes flying in all directions. Some clubs keep a frying pan handy.

Within a year of Captain Barr's suggestion, the NAR established an egg lofting event. The rocketeer had to fly one or more fresh grade A large hen's eggs to as high an altitude as possible and return the egg capsule to the judges. If the egg was cracked, it was a disqualified flight. Later, an egg lofting duration event was established.

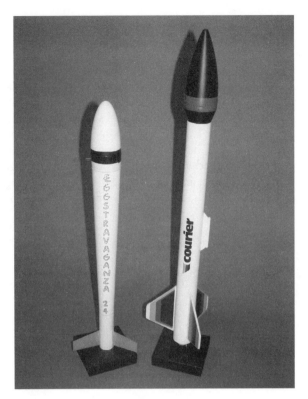

FIGURE 15-11 The Aerospace Specialty Products "Eggstravaganza 24" is designed for 24-mm rocket motors. On the right, the Quest "Courier" uses 18-mm motors.

Model rocketeers get very good at flying eggs. At one contest, I saw the same egg flown nineteen times by nineteen different model rocketeers, which certainly must stand as some sort of a record.

NASA later spent a couple of million dollars on a project that involved recovering fresh eggs by parachute. A lot of model rocketeers could have told NASA how to do it and used the money for more model rocket equipment!

There's always great excitement in an egg lofting contest, even when things don't work right. Egg lofting certainly separates the good model rocketeers from the bragging balsa butchers. If you goof, you have a hilarious mess on your hands. If you succeed, you can cook your payload for supper.

An offshoot of egg lofting is water lofting. This came about because a 2.7-ounce egg payload is no big deal for model rockets using Type F and Type G motors. The objective is to launch 8 ounces (a half-pint or a half-pound) of water to altitude and get the water back without losing any of it. This was quickly achieved because model rocketeers discovered that a plastic baby bottle would fit inside a common 2.6-inch-diameter model rocket body tube.

Water lofting is popular in hot, dry climates where you can drink your payload after you've flown it.

What else can be flown as a payload in a model rocket?

Just about anything that will keep the model's takeoff weight below the FAA exempt limit. There's not much to be gained by flying miniature kitchen sinks from doll houses. But many things could be flown in model rockets that haven't yet been aloft under rocket power. And some of these could be fun.

Payload-carrying model rockets present some interesting design and construction problems. They can be complicated and very expensive, depending on the kind of payload carried and the performance desired. In many cases, the model rocket is merely the inexpensive carrier for a very expensive payload. This is where the high reliability of model rocketry really pays off.

16

SCALE MODELS

Since model rocketry is space rocketry in miniature, it's only natural that model rocketeers would want to make their models resemble the big ones as closely as possible. The construction and flight of exact miniature replicas of full-size rocket vehicles and space vehicles—scale modeling—is model rocketry at its very best. It's the province of the true craftsperson.

One of the first model rockets ever built was a scale model. Chuck Moser, an engineer on the Pogo-Hi project at the Physical Sciences Laboratory of New Mexico State University, completed and flew his scale model of the U.S. Navy's Pogo-Hi parachute target rocket in March 1957.

Scale model rocketry combines craftsmanship with research. It reaches its pinnacle when a miniature replica, complete and correct down to the smallest details of marking and coloring, thunders off the launch pad for a straight flight and a perfect recovery. Months of research and work have probably gone into the model. Often such models are so beautiful that you hate to fly them. But having a beautiful scale model that's actually *flown* is much more satisfying than having one that sits forever earthbound on a shelf without ever being borne aloft under rocket power.

Scale models have been built with such fine workmanship and attention to detail that they qualify in all respects as museum scale models while also being capable of flight. Some model rocketeers even duplicate the launching facilities and equipment so that their scale model rockets take off from scale launch pads.

Many flying scale model rockets are in the collection of the National Air and Space Museum of the Smithsonian Institution. In some cases, they're the only models of a particular rocket vehicle that exist to help trace humankind's steps to the stars. As a result, they've become valuable additions to the national collection. The National Air and Space Museum has always been sympathetic to scale model rocketeers, some of whom have become the best astronautic historians alive today.

FIGURE 16-1   The epitome of model rocketry is the construction and flight of a scale model that duplicates in miniature a full-size rocket vehicle. Here George Gassaway prepares his radio-controlled scale shuttle model at the 14th World Spacemodeling Championships. The flight profile of George's model was exactly as the real shuttle with the SRB separating during boost (via an RC signal), then the liquid booster (also via an RC signal), and finally the shuttle itself returning via radio control. The flight was truly magnificent!

Scale model rocketry isn't something that can be done overnight. The creation of a scale model often takes weeks and months. It may take a long time simply to acquire the information that will ensure you really have a scale replica. And construction cannot be rushed. If you hurry, you're likely to put on that last frantic coat of paint on a perfectly built scale model only to have the paint blush or run, spoiling weeks of work. Don't start a scale model the night before a contest. Scale model rocketry takes weeks or months of planning, thought, careful workmanship, patience, attention to the smallest detail, and a lot of flight experience with nonscale sporting and contest models. Unlike participants in scale model aviation, scale model rocketeers are usually top-notch model fliers in their own right in addition to being scale modelers.

The rewards of scale modeling are great. You get an indescribable sense of pride and accomplishment when your model is placed on display after a flight. In contests, trophies and prizes are there to be won. In national and international meets, there are always events for scale models because scale modeling brings out the best craftspeople, designers, researchers, and contest fliers.

Once you've become a scale modeler, all other model rockets seem to be just paper and balsa look-alikes. The challenge of scale modeling never ends.

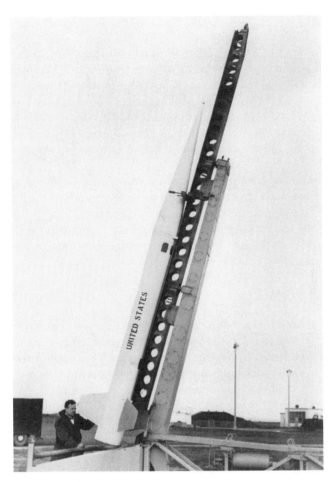

FIGURE 16-2 The real thing. A Nike-Smoke on the launch pad at NASA Wallops Flight Center. (NASA photo.)

When you think you've done everything, you'll always find another scale model beckoning to you. No modeler has ever built the perfect scale model. There are always improvements to be made, little flaws to be corrected on the next one, or there are new techniques to try. New full-size prototypes are always being developed, and these are fair game for the scale modeler. Some of the newer full-scale space vehicles, such as the NASA space shuttle with its complex shape and tile-covered Orbiter, pose real challenges to a person's modeling abilities.

By participating in scale model rocketry, you're likely to become such an expert in one particular space rocket or vehicle that you could talk intelligently with the project manager. And you'll discover that you've become hooked on the pursuit of excellence and perfection.

SPORT SCALE

You don't have to be a superexpert and top-notch craftsperson to get into scale modeling. You can get a good start in sport scale. All you have to do with a sport scale model is to build one that *looks like* the real one. The dimensions don't have to be precisely accurate. You don't have to have factory drawings or dimensions to prove your model is scale. As pointed out, it just has to look like the real thing. So the fins can be a little oversized, or the nose cone can be slightly short or long. The important factors are craftsmanship and appearance.

A sport scale model is judged from a distance of 1 meter (39 inches). The scale judges look for craftsmanship and the resemblance of the model to the real thing. And the sport scale model must make a safe flight.

Most of the factors discussed in this chapter apply to sport scale as well as to scale. Apply everything you can to your sport scale model and be prepared to move up to a good scale model once you've caught on.

IMPORTANT POINTS

You should follow several steps to build a good scale model. Occasionally, you'll discover an exception to these rules. However, keep in mind that these rules have been distilled from years of scale model work and my own personal experience in local, national, and international contests. Even if you don't wish to fly your scale model in contests, these points will serve you well and help you to achieve a perfect flying scale model of the prototype you like:

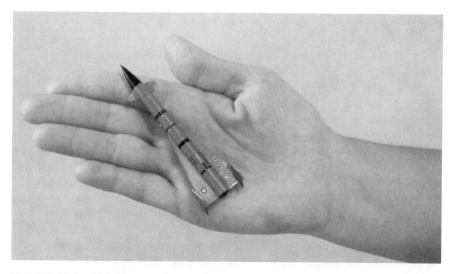

FIGURE 16-3   This "peanut" sport scale Jayhawk flies on a Quest micromotor. (Kit available from Aerospace Specialty Products.)

1. Select a good prototype to model.
2. Get adequate scale data before you start to build.
3. Using your data and model rocket manufacturers' catalogs, select the correct size for your scale model, using as many commercial parts as possible. Also select the type of model rocket motor that you'll use to propel your scale model.
4. Prepare accurate working drawings of the scale model and make accurate calculations of estimated CG, CP, weights, and flight performance.
5. Build a less-than-perfect flight test model first to ensure that your later, highly detailed scale model will perform as you want it to. You'll also discover and eliminate some of the construction problems you're sure to run into.
6. Build two or three detailed scale models of the same prototype side by side at the same time.
7. Take your time, do your best work, and don't take shortcuts.
8. Flight-test your less-than-perfect test model, make any necessary changes or alterations, and always fly your detailed scale model before the contest.
9. Keep building progressively better scale models of the same prototype.

SELECTING A SCALE MODEL

The most critical and important step in scale modeling is the selection of the proper prototype to model. If you don't do this carefully, you may have considerable trouble in building the model or getting it to fly. Too many modelers have become discouraged about scale modeling because they made the wrong choice of a prototype. As a result, they ran into insurmountable difficulties with the project.

The first step in the selection process is a self-evaluation. You must honestly determine your own abilities. If you're not very good at construction, assembly, painting, or other workmanship factors, you should select a prototype whose scale model will not be beyond your capabilities in your weakest area. Incidentally, this also holds true for *all* model rocket activities but even more so in scale model rocketry. You shouldn't select a complex prototype to model if you don't think you can finish it. This is not copping out on a challenge but simply being mature enough to recognize your own shortcomings and strengths in that stage of your development as a model rocketeer.

It's smart to begin with a simple model that will result in a good, reliable, flyable scale bird. Don't let your personal enthusiasm for a certain full-size rocket vehicle get the upper hand over common sense. You may be very eager to build a Saturn IB, a Titan-IIIC, or a space shuttle, because you've seen one launched, photographed one in a museum, or otherwise become taken with it. Or you may have been able to get all sorts of data on the old Nike-Hercules that's on display in the city park or the Atlas-D that's standing at the entrance of the nearby Air Force base. You may even have been able to get near a prototype to photograph it from all angles and measure it carefully. But do you really have the ability and experience to build a complex, highly detailed scale model on your first attempt?

The best model to start with is a simple, straightforward scaler that looks like the sporting model rockets you've been building. Your first scale model should be a single-stage vehicle with a cylindrical body, a simple ogive nose cone, and plenty of fin area. The color scheme should be simple and the prototype shouldn't have a lot of detail on it.

Another important selection factor that we'll discuss in greater detail later is the availability of information on the prototype. Scale information on current military rockets is usually very difficult, if not impossible, to obtain because of military security precautions. This is particularly true of recent guided missiles. It may even be true of many historic rockets and guided missiles as well, because the actual drawings and dimensions themselves may remain classified information even though there are hundreds of the rockets on public display all over the world. (It's also surprising what can be found in junkyards. I once spent $10 in a junkyard for a Nike rocket motor that I *knew* was classified. A friend of mine collected enough parts to build a *complete* Titan-II ICBM from dozens of junkyards from Boston to Baltimore.)

However, if you've chosen a very early rocket or space vehicle, you may have difficulty locating information because, believe it or not, pictures and drawings are often destroyed or thrown away to clean out the files and make room for the paperwork on the *next* project. Accurate information on many early rockets is difficult to find, even if the rocket was well known. For example, I have literally traveled all over the United States and Europe obtaining precise data on the old German A.4 (V-2) rocket. I've kicked myself from time to time because I was at White Sands when the U.S. Army was flying V-2 rockets there, and I've had my hands and tools inside V-2s that were launched a few hours afterward. Thus, although I remember a lot of things about the V-2, I never bothered at the time to collect scale data. So now I have to search it out, photographing one from a particular angle in the White Sands missile display, getting other data in the Deutchesmuseum in Munich, learning about another aspect in the Kensington Science Museum in London, and researching the library of the National Air and Space Museum. I know where *every* remaining German A.4 exists in the world. The only version I haven't seen is the Korolev V2B in the Kosmos Pavilion in Moscow, but I've got photos of it! I've worked years to accumulate all my scale information on the German A.4 and more than a hundred other rockets. It isn't easy doing the historical research for these rare birds.

For those modelers who are starting in scale modeling, I've found it best to begin with a kit. If built to sound scale model standards, these kits will give you a good taste of scale models, will start you toward learning the competent construction and finishing required for scale models, and will help you produce a good-looking scale model capable of holding its own in most competition.

There are several scale model kits available as of this writing. I hope there will be more available in the future. Three kits are currently on the market that can be made into good scale models that are easy to build and fly and that can get a modeler started in this fascinating field. The Quest Nike-Smoke, kit no. 3502, is an excellent scale model to begin with, even though much of it is prefabricated plastic. The Estes Black Brant II, kit no. EST 1958, is another simple scaler. For those who would like to build a big scale model, get the AeroTech I.Q.S.Y. Tomahawk, product no. 89015.

Other good scale model kits on the market, some of which are complex, advanced, and/or expensive are Estes no. EST 1380 Phoenix, Estes EST 1958 Black Brant II, EST 1952-V2, EST 2167 Mercury Redstone, AeroTech 89012 HV Arcas, and AeroTech 89015 Astrobee-D.

If you want to build from scratch, try your hand on these prototypes that make excellent scale models: USN Viking no. 10 (on display at the National Air and Space Museum), Sandia/NASA Nike-Tomahawk or NASA Nike-Cajun (on display at the NASA Goddard Space Flight Center), USN Pogo-Hi (on display at White Sands Missile Range), or U.S. Army Jupiter-C (on display at the National Air and Space Museum, Cape Canaveral, and the NASA Johnson Space Center). The biggest collections of prototypes for photographing and measuring are at the National Air and Space Museum, Washington, D.C.; the U.S. Air Force Museum, Dayton, Ohio; the Cape Canaveral Air Force Base Museum and the NASA Visitor's Center, Cape Canaveral, Florida; NASA Johnson Space Center, Houston, Texas; the Alabama Space and Rocket Center, Huntsville, Alabama; and White Sands Missile Range, New Mexico. Other smaller collections exist elsewhere.

When selecting a prototype, take into account these hints and tips concerning certain rocket vehicles:

The U.S. Air Force and U.S. Navy Aerobee series should be avoided because of the booster rocket, which is attached below the vehicle by an open tubular framework, making it a difficult model to stage. In fact, multistage scale models shouldn't be attempted until you have a lot more experience, because they're difficult and because there aren't that many good multistage prototypes around.

Air-launched missiles, such as the U.S. Air Force Falcon series and the U.S. Navy Bullpup, Sparrow, and Sidewinder missiles, should be avoided because they have guidance-control fins up near the nose or were designed to be launched from airplanes flying at high speeds.

Anti-aircraft rockets and similar missiles, typified by the U.S. Army Nike series, the U.S. Navy Tartar, Terrier, and Talos series, and the Soviet SAM missiles, should be tackled only by experienced scale modelers.

The Thor, Jupiter, Atlas, Titan, and Minuteman missiles are finless and therefore require the addition of clear plastic fins to make them aerodynamically stable scale model rockets. They should be attempted only by experienced scale modelers.

Unless you're *really* very good indeed, stay away from the NASA space shuttle. It is *very* complex structurally, aerodynamically, and in appearance because of details such as the thermal protection tiles.

Even with these caveats, there are hundreds of rockets and space vehicles suitable for scale modeling, and new ones turn up all the time.

OBTAINING SCALE DATA

Once you've chosen a prototype to model—and choose two or three different ones just in case you run into difficulty at some later stage—the next step is to get information that will permit you to build a true scale replica and not just a semiscale bird that looks something like the real one. (On the other hand, only a good photograph is required for sport scale.)

You may have already done some research in the local library, which may have many books with pictures and some very rough drawings. The NAR has a growing series of scale plans available to members (another enticement to join). A good source is a fellow model rocketeer who's collected photographs and drawings. As a matter of fact, a great deal of scale data swapping goes on in the NAR among scale modelers.

Peter Alway has published a compendium of scale data called *Rockets of The World* (available from Saturn Press). This book is a scale modeler's dream! Mike Dorffler Replicas also has a great CD collection of scale drawings of many Russian rockets as well as NASA Saturn 1B and Saturn V drawings.

However, to get really authentic data, you should go to the source—the manufacturer who made the prototype. If you don't know who made it, find out who uses it, such as NASA or the Air Force. Send your first letter to the user unless you know who the manufacturer is.

Your letters should be neat. Don't scrawl a penciled note on a scrap of notebook paper. Include your *complete* return address, including your zip code. The results you get may depend on the appearance, neatness, correct spelling, and correct grammar of your letter.

Don't concoct a fake aerospace research center letterhead or give yourself a fancy title. This doesn't help you at all and will, in fact, hurt your request. Everyone in the aerospace business knows everyone else and also knows what's going on in the business. A phony letterhead will be spotted immediately and will be chucked into the circular file with the crank mail and nut letters.

Your letter should clearly state that you are a model rocketeer who wants to build a scale model. You should ask for specific data such as a color photograph and a dimensioned external drawing of the vehicle. Be specific. Don't ask for all the information they've got. You won't get it even if they're getting ready to throw it away. Remember, no aerospace company or government agency is going to turn themselves inside out for you, taxpayer or youth science education notwithstanding. They'll probably send you whatever they're able to lay their hands on in a 5-minute scan of the public affairs office and files—and the public affairs office is the place where your letter will end up because no one else really knows what to do with it or has the authority to send you what you've requested. Besides, people in aerospace companies are always short of money and pretty busy, too. I'm not deriding aerospace companies or government agencies; some of them really will give you a lot of help, and they've helped other model rocketeers in the past. I'm simply predicting their most likely behavior toward you so that you won't get angry with them. Put yourself in their shoes, trying to respond to several letters like yours every day. Some companies and agencies are very good about sending data and photographs to model rocketeers, having anticipated your request and hundreds of others like it and gotten all the information together into modelers' packs.

It may take several weeks to get an answer. If you don't get one in four weeks, write another letter. Be polite. You're asking, and they're giving.

The result of your initial request may be nothing more than a full-color brochure that's beautiful and expensive but contains very little useful scale information. Write again, thanking them for the brochure and asking again for the specific information, explaining that you need it to build an exact scale model of *their* product that will eventually have *their name* on it for everyone in town to see.

Above all, don't give up! Persistent, polite, intelligent model rocketeers have succeeded in getting valuable historical information declassified or rescued from eventual repose in the trash can. They have performed a highly commendable service to aerospace history.

Remember, it's *quality,* not *quantity,* of information that you want and need. Don't ask for the entire stack of factory drawings. There may be thousands of these amounting to thousands of pounds of paper, including details of every nut and bolt that went into the prototype. You don't need all that, and it's actually too much data.

MINIMUM SCALE DATA

Although you may not be entering your first scale models in contests, it's a good idea to follow competition rules from the very start when it comes to acquiring minimum scale data. The consensus of the best scale modelers is represented in the NAR scale model rules, and almost anyone can obtain the NAR minimum scale data. In fact, I don't know how a scale model could be built with less than the required minimum scale data, which consist of

1. Scale factor (more about this in a moment)
2. Overall length
3. Diameter(s)
4. Nose length
5. Fin length, width, and thickness, if the prototype has fins
6. Length of transition pieces, if the prototype has them
7. Color pattern documented either in writing or by photographs
8. One clear photograph, halftone reproduction, or photo reproduction
9. For all dimensioned data listed above, both the actual prototype dimensions and the scale dimensions used on the model, which must be presented either on a drawing or in tabular form

Additional data are certainly desirable.

It's sometimes possible to measure or tape out a prototype if you can get near it. If you do this, you should make careful notes as you go along, then date the notes. A camera is an absolute necessity for recording details and shapes, and color film should be used for documenting color data. You might be surprised at the number of prototypes on display, just waiting to be taped out.

Don't wait to do this. Most rockets and guided missiles that are on display, especially outdoors, won't last forever. They weren't designed or built to last forever. They were built to be shipped and shot. For example, some of the early manned space capsules are disintegrating because they were designed for only one flight; even inside a museum, dissimilar metal corrosion or corrosion due to humidity is slowly eating them, reducing them to a pile of powder. You may end up having the only measurements and photographs of a display that everyone took for granted for years.

Scale data can sometimes be obtained from a professional or museum model, but don't count on the data being accurate unless you can back it up from an independent source. Some models aren't accurate, even museum models. In spite of advertising claims, very few commercial, nonflying plastic model kits are accurate. Remember that the museum model builder or the plastic kit manufacturer has been faced with the same scale data acquisition problems as you. And because of deadlines, the builder has usually had to proceed using incomplete and even inaccurate data.

You should begin to collect and file scale data because their quantity soon begins to grow. Scale data become valuable not only in building models but for trading with other scale buffs for data that they have and you need. Make a file folder or get a large manila envelope to hold all the information about a given prototype. Clip magazines and newspapers. Use copying machines to get copies of the data you want. Scale data need not be original documents because the original documents may no longer exist. Good-quality copies are useful and acceptable. Remember that you want and need quality data, not sheer quantity.

As you continue to collect scale data, you'll discover that you can't trust everything. Data from different sources may not agree. You may have to do some careful research to determine which data source is correct. You'll run into the problem of the lost inch where a mistake was once transferred from one drawing to another without anyone catching it. This may go on for years before someone, usually a scale buff, finally catches it. In some cases, I can spot exactly what data source was used by a modeler because of mistakes in the data that have shown up in other data and additional research. I also know when somebody uses some of my early scale drawings by recognizing the mistakes that were in those drawings, although I'd used the best available data at the time.

You should build a model of a particular prototype with a specific serial number, flight number, paint pattern, etc., unless the prototype was manufactured by the thousands on an assembly line with little or nothing except a stenciled serial number to differentiate individual vehicles. If the prototype wasn't mass produced, make a model of a particular vehicle, because many variations can occur in individual versions of a prototype. For example, every one of the

14 U.S. Navy Viking rockets launched was different. Over 65 percent of the German V-2 rockets launched at White Sands had major external changes in their basic configuration, and every one of them had a different paint pattern. The Freedom-7 Mercury capsule was significantly different from the Liberty Bell-7, and if you build a Mercury-Redstone, you'd better have the proper Mercury capsule atop the properly marked and numbered Redstone booster. Most of the Saturn-I vehicles were different, and the paint pattern usually shown for the Saturn-V was not the one on the actual flight vehicles. If you get involved with a launch vehicle such as Delta or Ariane where a lot of them have been launched, no two looking alike, you've got a lot of research to do.

DESIGNING A SCALE MODEL

When you've collected enough scale data to get started, you're ready to begin the design phase of your scale model, unless you're building from a kit. You should become completely familiar with the commercial model rocket parts that are available so that you can use them wherever possible. You should know what model rocket motors are available and what their performance characteristics are so that you can choose the proper one for your model. With all of this in mind, you can begin to size your scale model.

Sizing is a very important step. If you've chosen a prototype with lots of external detail and many complex shapes, you should build a big scale model. Little details are extremely difficult to make and attach to a small model. And on a small model details tend to get lost because the human eye cannot see them as well. On the other hand, if you've selected a rather simple prototype, your model will look better if it's smaller. All things being equal, however, a large scale model will usually win over a small model of the same prototype.

You may also have to size your scale model on the basis of what you intend to do with it. If it's a pure competition bird, you don't have as many restrictions. But if you want a scale model that has good altitude performance, you need to look for lowest drag, high impulse-to-weight ratio, and a size and color scheme that will permit the model to be seen by altitude tracking crews.

Again, use as many commercial parts as you can. You'll have to make plenty of custom parts as it is, so don't make things difficult for yourself. Use commercial body tubes. Many other commercial parts can also be used or modified for scale models.

Match your scale model to an available motor before you start to design and build just as you'd match a sport model to a motor. Many scale models end up being too heavy for their propulsion.

No hard-and-fast rules exist for sizing a scale model because each case is different. Sometimes you have to do a little experimenting by building two or more semiscale flight test models of the same prototype in different scale to find out which one works best. I built the I.Q.S.Y. Tomahawk in four different scales with four different body tube sizes just to see which one flew best. They all flew equally well, but the little one gave the best altitude performance. I've also built the ASP in three different sizes for three different motor types, the Nike-Smoke in four different scales, and the NASA Shotput in three.

To design your scale model properly, you'll have to do a little engineering drawing or drafting. This isn't difficult, and you should learn how because it will help you in your other model rocket design activities. Drafting equipment is available today at reasonable prices. You'll need a good drawing board, a T square, a couple of triangles, a protractor, a curve or two, a drawing compass, and a good measuring scale. Get a little practice first by designing and drawing a couple of sport models. You'll be surprised how easy it is to do model rocket drafting. The availability of CAD (computer-aided drafting) programs for personal computers can eliminate the need to do actual drafting.

Having decided on the size of your scale model, you must now prepare a full-size working drawing of the model so that you can determine sizes and shapes of various parts. To do this, you must determine the scale of your model.

SCALING A MODEL

Scale refers to the relationship of a model's size to that of its full-size prototype. The scale of the model is the ratio between a dimension on the model and the same dimension on the prototype. The most common method of determining scale is to compare the diameter of the prototype to the diameter of the model.

For example, if the diameter of the prototype is 31.0 inches and you plan to use a body tube with an outside diameter of 0.976 inch, you divide 31.0 by 0.976 on your calculator. The result is 31.762295. Round it off because the number 31.762 is close enough. The scale of your model will then be 1 to 31.762, written as 1:31.762. It means that 1 inch on the model is equal to 31.762 inches on the prototype.

Using a calculator, divide every dimension of the prototype by the scale factor—31.762 in our example. This will give you the dimensions of every part of your scale model.

Naturally, a 30-degree angle on the prototype is still a 30-degree angle on the model, because an angle is not a dimension that changes with the size of the model. So don't divide angles by the scale factor, or you'll end up with a funny-looking model indeed.

Using the dimensions you've calculated, draw a full-size working plan of your scale model. A far easier and more accurate way is to take the prototype drawing to a blueprint or copying shop and have them make a photographic reduction or enlargement of the prototype plans to exactly the size of your scale model. Blueprint and copy shops work in terms of percentage enlargment or reduction. If the drawing of the prototype has a dimension *on the drawing* of 3.1 inches, for example, you've got a $\frac{1}{10}$-size drawing of the prototype, and you'll want the drawing reduced by the factor $100 - [(0.976 \div 3.1) \times 100]$ or 68.5 percent. Talk to them about it and explain what you want to do if you're confused. Better to work with them to get it right than to pay for a reduction that you can't use. Getting such a reduction may cost you a few dollars, but it will save you having to do a lot of drawing and will be more accurate.

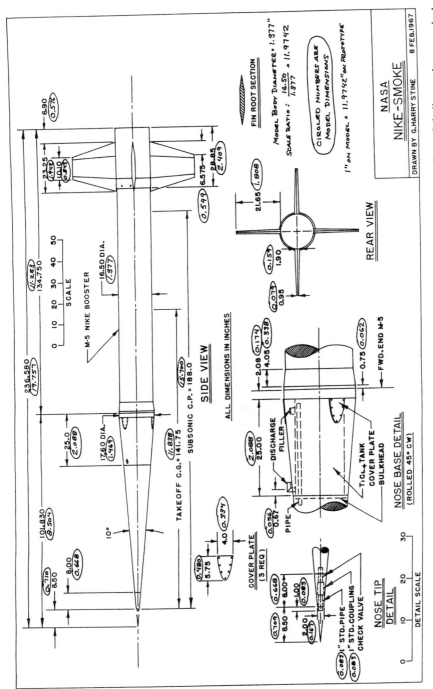

FIGURE 16-4   Scale drawing of the NASA Nike-Smoke rocket with both the prototype and model dimensions scaled properly.

FIGURE 16-5   A photograph of the real I.Q.S.Y. Tomahawk as launched at NASA Wallops Flight Center, Wallops Island, Virginia. (NASA photo.)

From the scale drawing you now have, you can determine the size of the parts. By tracing the fin outline onto a piece of stiff paper, you can make a template for cutting out the fins to the proper size and shape. You can also make paper templates for the nose shape, transition shape, boat tail shape, and other shapes on the model. These templates help ensure that these parts are shaped correctly.

Don't start to build yet! Stay on the drawing board and resist the temptation to cut balsa until you've really got that model figured out.

Calculate the CP. Make a trial run estimating the CG using the weight of the various parts and their distances from a common point such as the nose tip. This is the same procedure as calculating the CP, except you use the weight instead of the area.

Calculate the flight performance using RASP-93 or Rocksim and assume several different values of C_d to bracket the altitude range to which your model will fly. This will also help you make the proper choice of the model rocket motor and time delay to use. Sometimes, you can obtain the actual C_d of the prototype that you can use in the altitude calculations of the scale model, because C_d doesn't change with size.

Why all this paperwork? So that you can discover ahead of time whether your scale model will be too heavy, underpowered, or require too much nose weight to make it stable. You may discover that if you built it it wouldn't fly in a stable manner no matter what you did. You may even discover that it's impossible to build the model in the first place! In any case, time spent in preliminary paperwork pays off in scale model rocketry just as it does in full-scale rocketry. This is particularly true if problems show up in the design process. It doesn't take much time or money to redo calculations or change a drawing; it takes far more time to correct a mistake after you've started building the hardware.

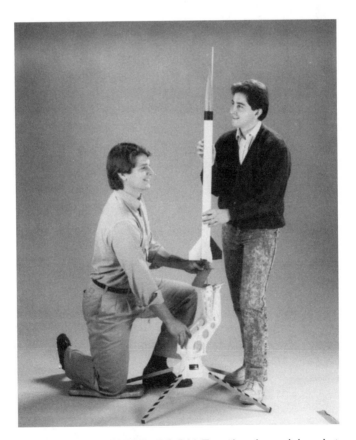

FIGURE 16-6   The scale I.Q.S.Y. Tomahawk model rocket available as a kit from AeroTech Consumer Aerospace. (Photo by AeroTech Consumer Aerospace.)

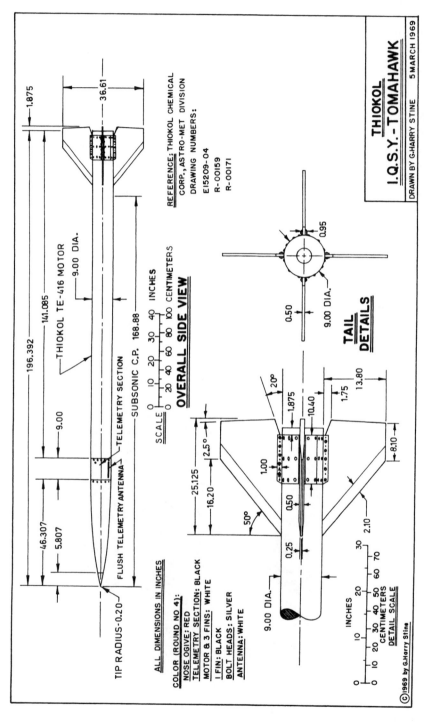

FIGURE 16-7   This scale drawing of the I.Q.S.Y. Tomahawk can be used to verify the scale of the AeroTech model or to build a smaller one. Permission is given to photocopy this drawing and the photo in Figure 16-5 for personal noncommercial nonsale use as contest scale substantiation data.

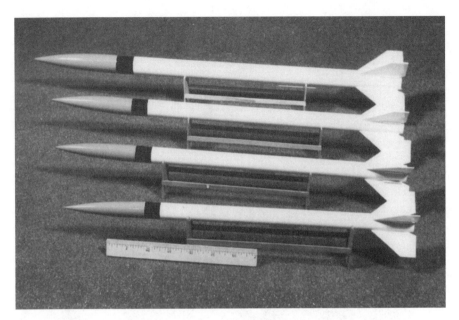

FIGURE 16-8 Building a scale model in a variety of different scales or sizes with body tubes of different diameters will help you get a good scale model that flies well. The author built these four I.Q.S.Y. Tomahawks in different scales to determine the best size.

BUILDING A SCALE MODEL

There aren't many special tips on building a scale model if you've done everything correctly up to this point and are a careful model builder. Just take care and do your best work.

Build a less-than-perfect semiscale flight test model first. Do a good job, but don't superdetail it. This is strictly a test bird. Build it and use it to check your calculations, to see that it flies as anticipated, and to correct any little operational problems that you may discover by doing all this.

When you finally sit down to build the superdetailed scale model, build two of them at the same time. It doesn't take much more time and effort to make a pair, and it will give you a spare. Somebody may sit on your scale model minutes before you get ready to fly it in that big regional contest.

You might also go back and bring your semiscale flight test model up to full scale-model standards, giving you three scale models of the same prototype.

IMPROVING YOUR SCALE MODEL

Most NAR champion scale modelers improve their scale models by building the same model over and over in an increasingly improved fashion. Every time you build a scale model, you learn tricks and shortcuts you can use the next

FIGURE 16-9   Except for size, the launch of a scale model can be as exciting as the real thing. John Langford watches as his NASA Javelin lifts off. (Photo by Alan Williams.)

time you build the same one. You also improve your modeling technique and skill each time. Not only do your scale models get better, but you can build them quicker.

J. Talley Guill, many times the U.S. junior and leader national champion, built models of the USAF-Convair MX-774 HiRoc for over five years, perfecting them to the point where they flew exactly as he wanted. Charles

Duelfer built models of the USAF GAR-11 Falcon with continual improvement. Bob Biedron kept building and improving his scale model of the French Ariane space launcher until he finally won first place at a world championship in 1992. Otakar Saffek of Czechoslovakia, the 1972 and 1974 world champion scale modeler, built over a dozen 1:100 scale Saturn V models, improving them each time. His final achievement was perfect down to scale corrugations, and it took him 2,000 hours of work.

Once you get a good combination, stick with it and improve it. Do research on additional details. Lighten and strengthen the model. Strive for high flight reliability. When you get tired of one scale model, put it away and try others, then come back to it later.

When you tire of building scale model rockets, try a scale launch complex. These are undoubtedly the most impressive scale models in the world. Often, they have remotely operated motorized moving parts such as launch rails that move up and down and turn in azimuth just like the prototypes. Others have service towers and gantries that move away under remote control. Some have umbilical towers and cables that swing away just before ignition. Special effects experts from Hollywood have nothing on our advanced scale-model rocketeers, some of whom have actually gone to work in Hollywood. Rick Sternbach, an outstanding model rocketeer, became chief illustrator and technical consultant for *Star Trek: The Next Generation.*

In scale modeling, there is literally no end to what you can do in miniature. It can give you a lifetime of enjoyment and challenge. When you're a good scale modeler, you're one of a very small and elite clan of model rocketeers.

We need more good, elite model rocketeers. Come on in! There's plenty of room for more people in scale model rocketry.

17

ALTITUDE DETERMINATION

Earlier we discussed methods of calculating the altitude performance of model rockets and learned that we can't accurately determine achieved altitude by calculation alone. We cannot control or accurately measure all the parameters and variables involved. Therefore, the only way to find out how high a model rocket will go is to actually fly it and use one of several methods for measuring its altitude.

We can't determine the achieved altitude of a model rocket by timing its flight to apogee and assuming zero-drag characteristics. This should be patently obvious in light of our earlier discussion and calculations of aerodynamic drag effects.

The electronic altimeters mentioned in chapter 15 are very accurate (some within 1 foot of actual altitude). The following are other methods that can be used with great success.

MARKER STREAMER ALTITUDE DETERMINATION

One altitude determination method has been developed that uses timing. It has proven itself to be reasonably accurate. At the suggestion of Douglas J. Malewicki, then an aeronautical engineer at Cessna Aircraft Company, and Larry Brown, a model rocket designer, Bill Stine in 1974 decided to look at the possibility of determining altitude by timing the fall of a standard marker streamer with a fixed set of dimensions and a fixed weight.

The theory behind this came from experiments conducted earlier by Malewicki and Brown dropping streamers from a known height on the Phoenix, Arizona, fire department's training tower. They discovered that a streamer would fall at a constant speed. They tested a series of streamers made of 0.0001-inch-thick polyethylene film 1 inch wide and 12 inches long with a 3-gram weight (or a penny) taped to one end. This standard marker streamer drops at a constant rate of 18 feet per second.

FIGURE 17-1 The Estes MaxTrax capsule is an electronic timer that times the fall of the capsule from apogee, then calculates the altitude.

Bill Stine decided this would make a good high school science fair project for him. If the standard marker streamer is packed into a model rocket atop the recovery device and ejected at or near apogee, the time required for the marker streamer to reach the ground will be equivalent to altitude at the moment of ejection. In a contest, the model whose marker streamer takes the longest time to reach the ground obviously ejected the marker at the highest altitude and therefore is the winner.

Some interesting facts emerged when Bill Stine flew his experimental flights. He recorded a spread in marker drop time of as much as 15 percent in identical models powered by same-type model rocket motors from the same production lot. He came to the conclusion that this variation in altitude was caused by random statistical variations in motor total impulse.

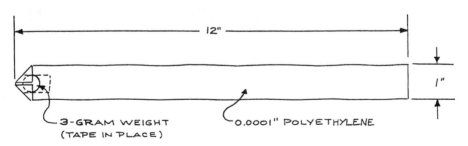

FIGURE 17-2 A sketch of the SAM (standard altitude marker) streamer developed and tested by Bill Stine.

Marker altitude competition remains an unofficial NAR contest event. However, it's perfect for schools and small clubs that don't have other altitude tracking systems and equipment. These altitude tracking systems may appear to be complex at first, but they are really very simple. And you don't have to know any mathematics other than arithmetic. When a big rocket flies at White Sands or Cape Canaveral, its flight performance and achieved altitude are determined by tracking the rocket in flight with electronic systems such as radar or with optical devices like telescopes or laser rangers (lidar).

In model rocketry, tracking is also used. However, electronic tracking of model rockets hasn't been used to date because of the high cost of equipment and the small size of the models. Although model rockets have been tracked by radar at NASA Wallops Flight Center, such electronic tracking appears at this time to be impractical because of costs and complexity. But progress in miniaturized electronics could change this very quickly.

Optical tracking is almost universally used in model rocketry. It is a fancy term for following the model in flight by eye and aiming some sort of measuring device at it. Without a lot of expense and knowledge, simple and reasonably accurate optical tracking equipment can be built by almost anyone with access to a junior high school shop or a well-equipped home workshop.

Optical tracking provides only one piece of information: a figure for maximum altitude achieved. This is an important datum, however, because it's use-

FIGURE 17-3   Model rockets can be tracked in flight with very simple equipment to obtain a definite figure for achieved altitude. Here, in national competition, two trackers are used at a station, one as a backup for the other.

ful in design evaluation, staging studies, and competition. Optical tracking combined with simple trigonometry—simple because you don't have to understand trig to make it work for you—has provided an easy, accurate means of obtaining this information since 1958. Three basic tracking methods are in use.

ELEVATION-ANGLE-ONLY ALTITUDE DETERMINATION

The first system uses a single tracking device that measures the elevation angle of the model as seen from a single tracking location at a known distance from the launch pad. Two people are required: one to launch the model, and the other to track it in flight.

An elevation-angle-only tracking device is shown in Figure 17-4. This is the simplest form of elevation-angle-only tracker. Quest Aerospace, Inc. sells an inexpensive unit called the SkyScope. Estes Industries, Inc., has a more expensive Altitrak unit. Both do the same job equally well.

The setup for this elevation-angle-only altitude determination method is shown in Figure 17-5. The launch pad is located at L, the tracker at T, and the model rocket at peak altitude at R. The distance LT is measured beforehand. This is called the *baseline*. For models that aren't expected to fly more than 500 feet high, a baseline distance of 300 feet (the length of a football field) is adequate; for models with higher performance, use a 600-foot baseline. The longer the baseline, the better the overall accuracy of the system.

FIGURE 17-4   The Quest SkyScope is an inexpensive and simple elevation-angle-only tracking device.

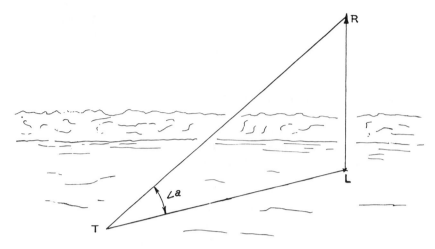

FIGURE 17-5   The geometry of the single-station elevation-angle-only altitude tracking scheme.

When the model rocket lifts off from L, we assume that it flies vertically over the launch pad, so the distance from a point on the ground directly under the flying rocket to the tracker doesn't change much from the measured LT distance. The tracker operator follows the model rocket visually to the peak of its flight, locks the tracking device with which he or she has been following the model, and reads the elevation angle that results.

The achieved altitude can now be calculated using simple trigonometry. We'll derive the equations here for the math buffs, but if you want to skip this, that's up to you. You don't have to know *why* technology works in order to make it work, although it certainly helps to know why when something goes wrong. (And it often does. Something about Murphy's Law.) It's not difficult to follow the derivations, and you might learn something in the process. Other model rocketeers have.

By definition, the angle RLT is a right angle. Therefore, according to the basic theorems of trigonometry, the *tangent* of the elevation angle $\angle a$ is equal to the achieved altitude divided by the ground distance LT as follows:

$$\tan\angle a = RL \div LT$$

Because we know the distance LT and the elevation angle $\angle a$, we can rearrange the equation, collect all the known factors on one side, and come up with

$$RL = LT \times \tan\angle a$$

The tangent of the elevation angle can be found by looking in a tangent table such as that in Table 8. You can find a tangent table in any high school trig text. The Quest SkyScope has the tangent tables printed on the side.

TABLE 8
Table of Tangents

Angle	Tangent	Angle	Tangent	Angle	Tangent
1	0.017	31	0.601	61	1.80
2	0.035	32	0.625	62	1.88
3	0.052	33	0.649	63	1.96
4	0.070	34	0.674	64	2.05
5	0.087	35	0.700	65	2.14
6	0.105	36	0.727	66	2.25
7	0.123	37	0.754	67	2.36
8	0.141	38	0.781	68	2.48
9	0.158	39	0.810	69	2.61
10	0.176	40	0.839	70	2.75
11	0.194	41	0.869	71	2.90
12	0.213	42	0.900	72	3.08
13	0.231	43	0.933	73	3.27
14	0.249	44	0.966	74	3.49
15	0.268	45	1.00	75	3.73
16	0.287	46	1.04	76	4.01
17	0.306	47	1.07	77	4.33
18	0.325	48	1.11	78	4.70
19	0.344	49	1.15	79	5.14
20	0.364	50	1.19	80	5.67
21	0.384	51	1.23	81	6.31
22	0.404	52	1.28	82	7.12
23	0.424	53	1.33	83	8.14
24	0.445	54	1.38	84	9.51
25	0.466	55	1.43	85	11.4
26	0.488	56	1.48	86	14.3
27	0.510	57	1.54	87	19.1
28	0.532	58	1.60	88	28.6
29	0.554	59	1.66	89	57.3
30	0.577	60	1.73	90	—

Let's run through an example of elevation-angle-only tracking. Suppose that the distance LT is 300 feet and for this particular hypothetical flight the elevation angle of the model rocket when it reached apogee is 32 degrees. The problem is solved as follows:

$$\tan 32° = 0.625 \text{ (from trig table)}$$

$$RL = LT \times \tan\angle a$$
$$= 300 \times 0.625$$
$$= 187.5 \text{ feet}$$

Not a high flight.

The elevation-angle-only method assumes that the model flies vertically over the launch pad, so the known ground distance LT doesn't change. Only a few models will have this sort of flight behavior. Most will weathercock into the wind. If the model flies *toward* the tracking station, assuming the altitude remains the same as in vertical flight, the distance LT becomes less and the elevation angle will be greater. Thus, the reduced data will show that the model apparently flew to a higher altitude than was actually the case.

This shortcoming can be partly eliminated by using a very long baseline LT or by locating the tracker T so that it's *crosswind* from the launch pad L. Then, when the model rocket weathercocks into the wind, it doesn't fly toward or away from the tracker, and the distance LT doesn't change very much.

TWO-STATION ALT-AZIMUTH ALTITUDE DETERMINATION

To fully account for the fact that a model rocket may not always fly vertically, a tracking system must be able to give an achieved altitude figure *no matter where the launch pad is located with respect to the trackers and no matter where in relation to the tracking station the model rocket reaches apogee*. To meet these requirements for 99 percent of all model rocket flights, a system was developed in 1958 using two tracking devices, each capable of following the model in elevation *and* azimuth.

An azimuth angle is an angle measured in the horizontal plane. A tracking device that will measure both elevation and azimuth angles is called a *theodolite*. Many different types of theodolites have been built and used in model rocketry. Some of them are very simple while others have been complex.

The basic parts of a theodolite are shown in Figure 17-6. This assembly may be threaded for mounting on any sturdy camera tripod. It may be made from wood, metal, or plastic. Inexpensive plastic protractors are adequate for use on such a theodolite, provided reasonable care is exercised to ensure accurate assembly. The elevation and azimuth axes must pass through the center of the protractors. Such a basic theodolite will measure horizontal azimuth angle and vertical elevation angle to the NAR's national standards of ±0.5 degree.

Several sorts of optical tracking aids may be attached to this basic theodolite base. The simplest and most effective is a straight piece of wood or metal 12 to 18 inches long with a headless brad driven into both ends. The brads provide two reference points with which to line up the tracker with the tracked object.

This sort of open bead sight with no optical lenses or magnification has been found to be excellent and highly accurate as long as the model rocket, its smoke trail, or a puff of tracking powder at ejection can be seen with the unaided eye. Telescopes or rifle scopes provide a restricted field of view, and high magnifications actually work to the detriment of tracking. If a model rocket cannot be seen in the first place, telescopes aren't going to help.

A somewhat more elegant sighting device is a mailing or body tube 2 inches in diameter and about 18 inches long with black thread crosshairs glued across both ends *and* two headless brads mounted atop the tube to pro-

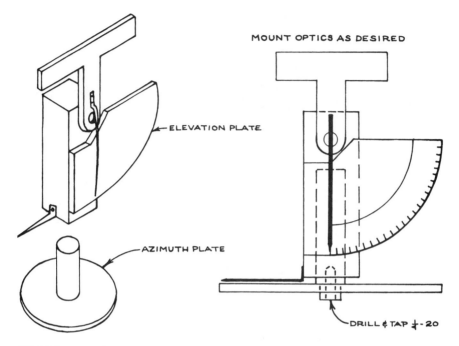

MOUNT OPTICS AS DESIRED

ELEVATION PLATE

AZIMUTH PLATE

DRILL & TAP ¼-20

FIGURE 17-6   Simplified sketch of an alt-azimuth tracking theodolite base that can be made in most workshops.

vide an alternate open field of view. The brads are used during the initial flight tracking while the crosshairs inside the tube permit accurate alignment of the theodolite with the model as it slows down near apogee.

An outstanding yet simple theodolite known as the Triple-Track Tracker, developed by Trip Barber, is shown in Figure 17-7. It can be made with common wood shop tools; instructions are provided in Appendix V.

The secret of tracking with this simple equipment is to watch the model with both eyes open, looking generally over the top of the tracker. Follow the model as it ascends, keeping the tracker moving with the model. But don't try to keep the model zeroed in the crosshairs at this stage of flight because you don't need to. You just want to keep it in sight. As the model slows down near apogee, *then* zero in on it with the theodolite and get it right atop the brads or right in the crosshairs. Model rockets $\frac{1}{2}$ inch in diameter and 8 inches long have been tracked to altitudes of more than 1,800 feet with this sort of equipment using no telescopes at all.

If a model rocket can't be tracked, the competition rules imply that it's not the fault of the trackers but of the builder of the model. This assumes that the builder built a bird that *couldn't* be tracked or painted it wrong. In science and technology, you're out of luck if you build something whose performance exceeds the capabilities of the measuring equipment. Better learn that now.

FIGURE 17-7   The easy-to-build Triple-Track Tracker designed and perfected by Arthur H. "Trip" Barber.

The two-station alt-azimuth system—as it's known—requires two trackers positioned at the ends of a measured baseline. The tracking situation is shown in Figure 17-8. For national and international competition the baseline must be at least 300 meters (984 feet 3 inches) long. Two tracking theodolites and the baseline may be set up in any relationship to the location of the launch pad. It is not necessary, and, in fact, *it is highly undesirable* to have the launch pad located on the baseline. Research on optimum relationships between the baseline and the launch pad has been carried out by J. Talley Guill in 1972 and Wally Etzel in 1993, and these results have been confirmed by my own computer analysis. The optimum location of the launch pad for best tracking accuracy is at an azimuth angle of between 30 degrees and 45 degrees from the baseline as seen from each tracking station.

The two tracking stations must have a clear view of the launch pad and of each other. The baseline between them should be horizontal—don't put one station atop a hill and the other one in a valley.

For the best view of the model in flight, position the stations and the baseline *south* of the launch pad. One station should be *southeast* of the pad and the other *southwest*. These locations permit both stations to see the model in the full light of the sun. Don't position the tracking stations so that one of them looks into the sun while tracking a model.

With the baseline measured and the stations located, the theodolite is set up at each station. Each theodolite should be leveled so that the azimuth pro-

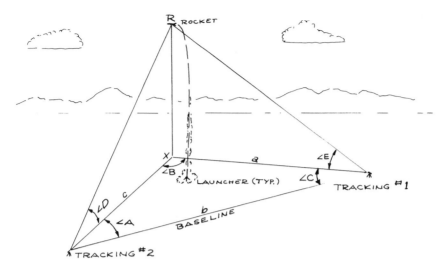

FIGURE 17-8   The geometry of the two-station alt-azimuth altitude tracking scheme.

tractor is indeed horizontal. A bubble level is often mounted on the azimuth protractor for this purpose. Once both theodolites are leveled, the system is *zeroed* by sighting *both* theodolites *directly at each other* along the baseline. When the theodolites are leveled and zeroed, their azimuth and elevation dials or pointers are set to read *zero* azimuth angle and *zero* elevation angle. Don't zero the theodolites on the launch pad. That doesn't work with this method. Zero them on one another.

We now have a tracking situation with a known distance between two stations plus an azimuth angle and an elevation angle from two theodolites once the flight is made. This is more than enough data to compute the achieved altitude.

To understand how this is done, let's derive the equations used, referring to Figure 17-8 in the process:

> *Given:* Distance b
> Angle ∠A
> Angle ∠D
> Angle ∠C
> Angle ∠E

Point X is an imaginary point on the ground directly beneath the apogee of the model in the air. The model rocket at apogee is at point R. We know the distance *b*; it's the measured baseline. So we need to compute the distance RX which is the vertical altitude of the model rocket above the ground.

If we compute the distance *a* or the distance *c*, we can theoretically locate point X on the ground. Then we can compute for the two vertical triangles to find the value of RX in two different calculations, checking one against the other to determine accuracy. These two vertical right triangles (R-X-West and R-X-East) are used to compute RX separately for each, and the value of RX from both is then averaged to produce a more accurate number.

The Law of Sines states

$$c \div \sin\angle C = b \div \sin\angle B = a \div \sin\angle A$$

Therefore,

$$c = \sin\angle C(b \div \sin\angle B)$$
$$= \sin\angle C\{b \div \sin[180 - (A + C)]\}$$

Since R is directly above X by definition, the angle R-X-West is a right angle. Therefore, we can compute the western triangle as follows:

$$\tan\angle D = RX \div c$$

Therefore,

$$RX = c \tan\angle D$$

Substituting for c, we get

$$RX = \sin\angle C \tan\angle D\{b \div \sin[180 - (A + C)]\}$$

The other vertical right triangle, the eastern one, is solved in an identical manner:

$$RX = \sin\angle A \tan\angle E\{b \div \sin[180 - (A + C)]\}$$

The two values of RX thus obtained are then averaged. If either value deviates by more than 10 percent from this average altitude, it can be assumed that something went wrong. It means "track lost." But if both values of RX come within plus or minus 10 percent of the average altitude, it's considered "track closed" and the achieved altitude data are considered to be official.

You don't have to memorize all this mathematical stuff to make the system work for you. All you have to do is be able to read some tables, write some numbers into blanks, and do some common arithmetic. We've worked out lots of simple ways to do it. Let's start with the most difficult and proceed to the simplest. Here's a tracking example:

> *Given:* Baseline $b = 300$ meters
> Tracking East azimuth ($\angle C$) = 23°
> Tracking East elevation ($\angle E$) = 36°
> Tracking West azimuth ($\angle A$) = 45°
> Tracking West elevation ($\angle D$) = 53°

For one triangle we compute

$$
\begin{aligned}
RX &= \sin\angle C \, \tan\angle D\{b \div \sin[180 - (A + C)]\} \\
&= \sin 23° \tan 53°\{300 \div \sin[180 - (45 + 23)]\} \\
&= 0.391 \times 1.327 \times (300 \div \sin 112°) \\
&= 0.391 \times 1.327 \times (300 \div \sin 68°) \\
&= 0.391 \times 1.327 \times (300 \div 0.927) \\
&= 0.391 \times 1.327 \times 324 \\
&= 168 \text{ meters (551 feet)}
\end{aligned}
$$

Don't be confused by the substitution of sin 68° for sin 112°. The sine of an obtuse angle (greater than 90 degrees) is by definition the sine of its supplementary angle (180 degrees minus the obtuse angle).

Solving the other triangle by the same method yields RX = 166 meters (545 feet).

The average of these two altitudes is (168 + 166) ÷ 2 or 167 meters, the average altitude.

In applying the 10 Percent Rule, the average altitude is rounded off using the rule of thumb, Keep it even—that is, if the average altitude ends in a decimal of 4 or less, the decimal is dropped; if it ends in a decimal number of 6 or more, it's rounded-up to the next whole number; if the decimal tenth is 5, it's rounded down if the number immediately to the left of the decimal is even and rounded-up if that number is odd.

Both computed altitudes—166 and 168 meters—fall within 10 percent of the average altitude. Therefore, the track is considered good.

This system was originally worked out in 1958 by two high school students, Arthur H. Ballah and Grant R. Gray. It was refined in 1960 by John S. Roe, who devised a table giving both sine and tangent functions *plus* a column for the value of 300 ÷ sin∠B. I revised it to work with any baseline. See Table 9. This requires only a pencil, paper, and a four-function pocket calculator.

The Universal Flight Data Sheet duplicated in Figure 17-9 is useful for entering your flight data and performing the calculations to determine the achieved average altitude using Table 9.

The next step was to put the whole system into a high-speed digital computer and instruct it to solve for every possible combination of azimuth and elevation angles. This was done by J. Talley Guill at Rice University in 1969 and produced the first of a series of precomputed altitude tables that were used at national meets for many years.

The introduction of the scientific programmable pocket calculator resulted in even simpler and faster data reduction without the need to do anything more than enter the angular data. The calculator/computer then spits out the two altitudes, the rounded average altitude, the percent error, and decides whether the track is lost. Gary Crowell was the first to write a program for this, and it was successfully used at NARAM-20 in Anaheim, California, in 1978. Wally Etzel and others refined the program for pocket calculators. Today, the achieved altitude can be announced before the model returns to the ground! The data reduction program in BASIC is shown in Appendix VI.

UNIVERSAL FLIGHT DATA SHEET   Date:_____

Name:_____ NAR#:_____

Model name:_____ Motor Type(s):_____

Launch Alley:_____ Safety Check by:_____

TRACKING DATA: Baseline Length:_____ Recorded by:_____

TRACKING EAST: Azimuth:_____° Elevation:_____° LOST □

TRACKING WEST: Azimuth:_____° Elevation:_____° LOST □

DATA REDUCTION: DR by:_____

(NOTE: If any angle MORE than 90°, subtract from 180° to get computational angle.)

EAST: Azimuth:_____° sin:_____ (1) Elevation:_____° tan:_____ (2)

WEST: Azimuth:_____° sin:_____ (3) Elevation:_____° tan:_____ (4)

Add azimuths for ∠B:_____ Table Value Q for ∠B:_____

MULTIPLY: Q × Baseline = _____ × _____ = _____ (5)

MULTIPLY: _____ (5) × _____ (2) × _____ (3) = _____

MULTIPLY: _____ (5) × _____ (4) × _____ (1) = _____

                                                    ADD: _____

THIS IS AVERAGE ALTITUDE→ DIVIDE BY 2:_____

10% RULE CALCULATION:

Average Altitude: _____ If RESULT is MORE than
                                           HIGHEST of the two

ADD 10% of average altitude: _____ computed altitudes, TRACK
                                           CLOSED.

RESULT: _____ □ CLOSED □ LOST

Average Altitude: _____ If RESULT is LESS than LOWEST
                                           of the two

SUBTRACT 10% of average altitude:_____ computed altitudes, TRACK
                                           CLOSED.

RESULT: _____ □ CLOSED □ LOST

FIGURE 17-9   The Universal Flight Data Sheet can be used with Table 9 for any baseline length. It is used to record and reduce angular tracking data if a programmable pocket calculator or laptop computer isn't available.

THREE-STATION ELEVATION-ANGLE-ONLY ALTITUDE DETERMINATION

A third tracking system was developed by J. Talley Guill, who believed it should be possible to have a tracking system using several elevation-angle-only trackers like the Quest SkyScope. He perfected his system in the summer of 1972. It was checked for accuracy by comparing it side by side with the

TABLE 9
Altitude Calculation Table for Two-Station ALT-Azimuth System

This table is for computing the altitude of models using azimuth and elevation angles from two tracking stations. To use with any baseline, multiply value Q by baseline length in meters or feet ($Q = 1 \div \sin\angle B$). See Figure 17-9, Universal Flight Data Sheet, for computational procedures.

Angle	Sine	Tangent	Value Q
1	0.0174	0.0174	57.298688
2	0.0349	0.0349	28.653718
3	0.0523	0.0524	19.107323
4	0.0698	0.0699	14.335587
5	0.0872	0.0875	11.473713
6	0.1045	0.1051	9.5667722
7	0.1219	0.1228	8.2055090
8	0.1392	0.1405	7.1852965
9	0.1564	0.1584	6.3924532
10	0.1736	0.1763	5.7587705
11	0.1908	0.1944	5.2408431
12	0.2079	0.2126	4.8097343
13	0.2249	0.2309	4.4454115
14	0.2419	0.2493	4.1335665
15	0.2588	0.2679	3.8637033
16	0.2756	0.2867	3.6279553
17	0.2924	0.3057	3.4203036
18	0.3090	0.3249	3.2360679
19	0.3256	0.3443	3.0715535
20	0.3420	0.3640	2.9238044
21	0.3548	0.3839	2.7904281
22	0.3746	0.4040	2.6694672
23	0.3907	0.4245	2.5593047
24	0.4067	0.4452	2.4585933
25	0.4226	0.4663	2.3662016
26	0.4384	0.4877	2.2811720
27	0.4540	0.5095	2.2026893
28	0.4695	0.5317	2.1300544
29	0.4848	0.5543	2.0626653
30	0.5000	0.5774	2.0000000
31	0.5150	0.6009	1.9416040
32	0.5299	0.6249	1.8870799
33	0.5446	0.6494	1.8360784
34	0.5592	0.6745	1.7882916
35	0.5736	0.7002	1.7434468
36	0.5878	0.7265	1.7013016
37	0.6018	0.7535	1.6616401
38	0.6157	0.7813	1.6242692
39	0.6293	0.8098	1.5890157
40	0.6428	0.8391	1.5557238
41	0.6561	0.8693	1.5242531
42	0.6691	0.9004	1.4944765
43	0.6820	0.9325	1.4662792

(continues)

TABLE 9
(continued)

Angle	Sine	Tangent	Value Q
44	0.6947	0.9657	1.4395565
45	0.7071	1.0000	1.4142136
46	0.7193	1.0355	1.3901636
47	0.7313	1.0723	1.3673275
48	0.7431	1.1106	1.3456327
49	0.7547	1.1504	1.3250129
50	0.7660	1.1918	1.3054073
51	0.7771	1.2349	1.2867596
52	0.7880	1.2799	1.2690182
53	0.7986	1.3270	1.2521357
54	0.8090	1.3764	1.2360680
55	0.8192	1.4281	1.2207746
56	0.8290	1.4826	1.2062179
57	0.8387	1.5399	1.1923633
58	0.8480	1.6003	1.1791784
59	0.8572	1.6643	1.1666334
60	0.8660	1.7321	1.1547005
61	0.8746	1.8040	1.1433541
62	0.8829	1.8807	1.1325701
63	0.8910	1.9626	1.1223262
64	0.8988	2.0503	1.1126019
65	0.9063	2.1445	1.1033779
66	0.9135	2.2460	1.0946363
67	0.9205	2.3558	1.0863604
68	0.9272	2.4751	1.0785347
69	0.9336	2.6051	1.0711450
70	0.9397	2.7475	1.0641778
71	0.9455	2.9042	1.0576207
72	0.9511	3.0777	1.0514622
73	0.9563	3.2709	1.0456918
74	0.9613	3.4874	1.0402994
75	0.9659	3.7320	1.0352762
76	0.9703	4.0108	1.0306136
77	0.9744	4.3315	1.0263041
78	0.9781	4.7046	1.0223406
79	0.9816	5.1445	1.0187167
80	0.9848	5.6713	1.0154266
81	0.9877	6.3138	1.0124651
82	0.9903	7.1154	1.0098276
83	0.9925	8.1443	1.0075098
84	0.9945	9.5144	1.0055083
85	0.9962	11.430	1.0038198
86	0.9976	14.301	1.0024419
87	0.9986	19.081	1.0013723
88	0.9994	28.636	1.0006095
89	0.9998	57.290	1.0001523
90	1.0000	∞	1.0000000

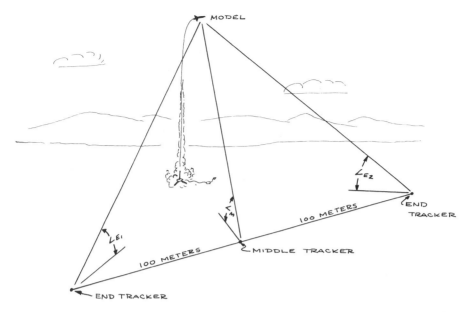

FIGURE 17-10   The geometry of the three-station elevation-angle-only tracking scheme. See Appendix VII for data reduction procedure and value tables.

two-station alt-azimuth system described previously. Three elevation-angle-
only trackers are required. They must be located on a single baseline as shown
in Figure 17-10. One tracker is at each end of the baseline and the third
tracker is precisely in the middle of the baseline. The data reduction tables for
the three-station elevation-angle-only system are given in Appendix VII. This
system works fine. It is probably the answer for those schools and clubs who
don't want to build theodolites and prefer to buy existing elevation-angle-only
trackers. At the time of this writing, no one has written a data reduction program for this system that can be run on a scientific pocket calculator. But I'm
sure they will, if I don't decide to do it some night in a fit of creativity.

COMMUNICATIONS WITH TRACKING STATIONS

Even when using a single elevation-angle-only tracker, you must have some
means of communication between the launcher and the trackers if for no other
reason than to inform the trackers of an imminent launch so that the liftoff
doesn't come as a complete surprise.

The simplest range communications system I ever saw was used at the
First International Model Rocket Competition in Dubnica-nad-Vahom,
Czechoslovakia, in May 1966. One handheld flag is at the launch pad and
others are at the tracking stations. When a model is ready to launch, the flag
person at range control raises the flag. The trackers see this and raise their
flags when ready to track the model. When the range-control flag person
drops his or her flag, the model is launched. The trackers then write down

their angles and send them back to range control by runners. (The Czechs used motorcycle couriers instead of runners.)

For more than thirty years, range communications between the trackers and the range-control center usually used hard-wire land-line telephone circuits. These required lots of twisted-pair wire, telephone handsets, and other accessories. Most model rocket clubs now use Family Radio Service (FRS) radios for range communications. FRS radios are widely available and require no special license to operate. They have a maximum power of $\frac{1}{2}$ watt and a range of up to 5 miles. Most FRS radios have up to twenty-two channels, enabling each range operation to use one dedicated channel (for example, trackers report on one channel, range safety another). A channel protocol for the rocket range is often posted at the RSO's table. FRS radios are also very handy for recovery of your rocket when it drifts far away from the launch site. When flying an HPR rocket or for a duration competition, a spotter (or several spotters) all on the same channel can be used to aid in keeping an eye on the rocket.

With your tracking and communication system, you're now well on your way to having a complete model rocket range on which you can hold regional competitions!

18

MODEL ROCKET RANGES

Although you may begin flying model rockets in an open field with one or two other people, you'll soon be joined by others who want to fly there, too. You can have a lot of fun and learn more when you fly with other model rocketeers. However, if many models are flown and if they are to be tracked for altitude, some additional equipment will be required. And you will need some form of organization to maintain safety and prevent confusion.

When most people visit a model rocket range for the first time, they're impressed by its organization and safety control. These model rocket range characteristics did not evolve in a haphazard manner. Model rocket range operations were carefully designed because *safety is no accident*.

The first time model rocketeers got together to fly, it was at White Sands Missile Range in New Mexico. We were all aware from our professional rocket work that range organization was absolutely necessary. We'd been thoroughly indoctrinated with the safety procedures used in flying real rockets and guided missiles. Few model rocketeers realize the tremendous debt they owe to two men who pioneered flight safety practices in full-scale rocketry and whose policies were adopted by model rocketeers. They are Herbert L. Karsch and Nathan Wagner, the flight safety engineers at White Sands back in the 1950s. Much of what you're going to read about in this chapter was adopted straight from White Sands, where all of us got our early training and experience, because at that time it was the *only* place to get it. It's largely because of Karsch and Wagner that model rocketry, since its inception in 1957, has conclusively proven itself to be safer than model airplane flying and safer even than bicycle riding.

The number one safety rule adopted from professional rocketry is: *The range safety officer's word is as the word of God*. No one can override a safety decision made by a range safety officer (RSO). If the RSO says no, that's the end of it. Either obey or go someplace else.

At White Sands, Cape Canaveral, Vandenberg Air Force Base, Point

Mugu, Eglin Air Force Base, Fort Churchill, Tonopah, Green River, and NASA Wallops Flight Center, the RSO is supreme. The RSO or the range safety crew checks every rocket vehicle before flight. The RSO alone determines whether the rocket is safe to launch. The RSO alone determines that the range is ready and in a safe condition. The RSO gives the final safety clearance before launch. The RSO even destroys the rocket in the air if it becomes unsafe.

The basic rocket range philosophy for the big ones and the little ones is: *Don't take chances.* It isn't worth it. Those of us who started model rocketry profited from the experiences of the professional range safety engineers and adopted as many of their practices as practical. The only thing we don't do in model rocketry is destroy the model in midair if it becomes unsafe (although I have often wished we had such a procedure). I once remarked on this to the former director of NASA Wallops Flight Center, Robert Krieger, who replied as we watched a model lift off, "Well, if you'll get me a twelve-gauge shotgun. . . ."

While adopting professional range safety procedures for model rocket ranges, we also borrowed many other aspects of full-scale rocket flight testing. Range operations procedures were one such aspect. Although model rocketry is a very individualistic hobby where a person can use creativity to the utmost within the bounds imposed by the real universe in the design and construction of model rockets, it's also a team effort on the model rocket range where everyone works together.

A model rocket range is a little Cape Canaveral. It's run the way the Cape is. If anyone who had a rocket to fly were allowed to set up a launch pad anywhere at the Cape and launch whenever it was convenient, it would be chaos. The same holds true for a model rocket range. It's not only fun to work together, but operating this way helps tie our hobby closer to the real thing.

The first requirement for a model rocket range is a large open space of land. For most model rockets flown under the NAR Model Rocket Safety Code, an ordinary school football field is adequate if it isn't hemmed in by too many rocket-eating trees. If you're going to fly large models or do a lot of glide-recovery flying, you'll need a larger field. Model rocketeers have the same problem as model aviators: finding a flying site large enough. In the western and rural parts of the United States, model rocketeers usually don't have too much of a problem finding a flying site if they're willing to drive ten or twenty miles. Someone always has a friend or relative with a farm or ranch land somewhere. But this isn't true if you're among the 90 percent of American people who live in urban and suburban areas. As the old maps used to point out, here there be complaining neighbors, uncooperative fire marshals, and/or insolent bureaucrats or politicians. But these people can usually be transformed into friends by careful diplomacy.

We have a wonderful square mile of land as a rocket range outside Phoenix, Arizona. One day after we'd been flying there for about eighteen months, the sheriff showed up. When asked what the problem was, he reported that some lady nearby had complained that we were "shooting Scud missiles at her." (Sigh.)

There are no pat answers to the problem of finding a place to fly. But you should always have the *written* permission of the owner of the land. In my own experience, it's been easier to get permission to use privately owned land such as a farm meadow. Getting permission to use public land usually means dealing with a bureaucrat somewhere, and this can become frustrating because the easiest and safest thing for such a person to say is no. However, clubs that are sponsored by local civic groups have had very good success in getting permission to use parks and recreational areas.

The most important single thing that will help a group of young model rocketeers obtain permission to use a flying site is an adult adviser who may be a member of the club and who's willing to act for you. An adult will know how to approach local authorities and can speak as a tax-paying, voting citizen. Believe me, this helps! An adult spokesperson can also serve as the club's range safety officer, club adviser, or club sponsor. More about that in the next chapter.

Any flying field that you select should be reasonably free of trees, of course. Regardless of what the botanists say, all trees are rocket-eating trees. The field should also be clear of high-voltage electrical transmission lines; any model that gets hung up in a power line must be left there because getting it out could be deadly.

A flying field should also be away from heavily traveled roads, highways, parkways, turnpikes, freeways, and interstates. A model rocket probably won't damage an automobile, but if a parachuting model wafts down in front of a driver speeding along in heavy traffic, it's certainly going to be startling and may cause an accident.

Stay away from airports, airport runway approach zones, and other areas where there may be low-flying aircraft. Model rockets are *perceived* as a possible danger to airplanes, although this perception is totally false. It's totally impossible to hit an airplane in flight with a model rocket. NAR studies have shown that the chance of a model rocket hitting an airplane in flight is one in 50 billion with 95 percent confidence level. This has been confirmed by actual tests conducted by the U.S. Army Ordnance Corps at White Sands. The Department of Defense has spent billions of dollars to develop guided missiles that will deliberately hit airplanes. A little unguided model rocket isn't going to achieve what an expensive, complex, and sophisticated Chaparral, Hawk, or Redeye SAM is designed to accomplish. If model rockets were really dangerous to airplanes, why did the Department of Defense spend billions of dollars on SAMs since 1945?

With the permission of the airport manager, flying model rockets directly on an airport itself is probably the safest place to do it in busy urban airspace. National model rocket competitions have been flown with perfect safety in the middle of active first-class airports (Lawrence G. Hanscom Field in Bedford, Massachusetts, for example).

The layout of an *idealized* model rocket range is shown in Figure 18-1. Note that I emphasized the word *idealized*. Someday I'd like to fly on such an ideal rocket range. I've been on some very good model rocket ranges, but

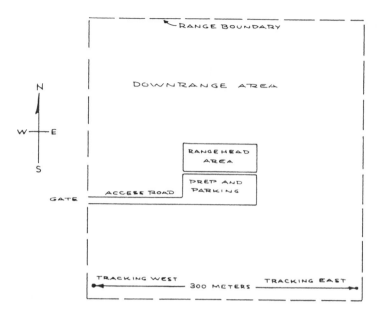

FIGURE 18-1   The layout of an idealized model rocket range.

every location is somewhat less than ideal. We do have a good one in the Phoenix, Arizona, area where we have the use of a square mile of state land.

Note that the launch area is located at or near the center of the field. This makes flying and recovery independent of the wind direction. However, if the prevailing wind is predominantly from one direction, you can offset the launch area on the *upwind* side of the field to provide more downwind recovery area.

The tracking stations are located on the south side of the field so that the trackers don't have to look into the sun when following models in flight. Note that the trackers have a clear and unobstructed view of the launch area.

The launch area or range head normally faces north so that the range safety officer (RSO), the range-control officer (RCO), and model rocketeers have the sun *at their backs* while flying. You may not think this is very important until you've been on the range all day long and discover your sunburned tonsils due to looking up into the sun. A couple of national meets were accidentally set up with the launch area facing into the sun, and this resulted in five miserable days of model rocket flying and some very painful sunburns.

Speaking of sun, please apply a sunblock before going to the range. Skin cancer isn't funny. Here in Arizona, we use the strongest we can get: 45.

There should be only one entrance to the flying field. A sign should be posted to let people know what's going on and to advise them to stay alert if they come on the field to watch. The wording of the following sign is recommended:

PODUNK MODEL ROCKET CLUB
FLYING FIELD
Model Rockets in Flight
PROCEED WITH CAUTION

The sign should not say such things as "Danger! Look Out for Falling Rockets!" or other such drastic warnings that suggest that you're not conducting a safe activity.

Cars should be parked so that they don't block the line of sight from the trackers to the launch area. If possible, cars *should* be allowed on the field because most model rocketeers work out of the trunk or the back end of a station wagon. I've been on fields where cars were not allowed, and we had to carry everything for more than a hundred yards into the launch area. It was miserable.

The launch area is usually called the *range-head area*. Figure 18-2 shows the layout of a typical modern range-head area.

The range-head area is separated from the spectator and prep area by a simple rope barricade. There is an open gate in the center. The rope barricade keeps spectators out of the range-head area. Everyone who enters and leaves the range-head area should use the gate and not jump the barricade. If it's a big range-head area, put in extra gates to keep modelers from jumping or ducking the barricade.

The barricade doesn't have to be sturdy. It's basically a symbolic barrier. Often a gasoline station or a used car lot will give you some of its used bannered ropes. These are colorful and attractive, and they liven up the appearance of the range. They are made into a barricade by using steel fence stakes driven into the ground to support them. A post should be located every 20 feet or so.

Fifty feet beyond the barricade is the line of launch pads. Each launch pad

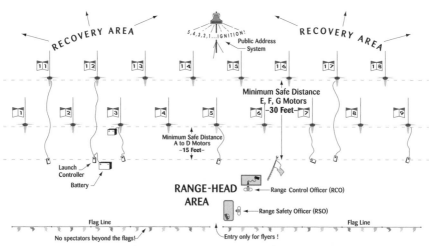

FIGURE 18-2 The layout of the range-head area of a model rocket range equipped to fly up to G motors. (Drawing by Ed LaCroix.)

site is marked by a sign with a number on it. An easy way to make launch pad markers is to buy several white plastic wastebaskets, turn them upside down, and paint a number on each.

Launch pads should be separated by 10 to 25 feet in the lineup. If you're flying on a field with dry grass, each launch pad site should have a canvas tarp at least 5 feet square on which the launch pad can be set up. The tarp serves two purposes. It keeps you from getting your pants or knees dirty when you're on your knees hooking up your model. It also prevents grass fires.

A grass fire is no fun. On March 31, 1967, our club accidentally burned off an 11-acre grass field. All safety precautions were in effect, but a glowing piece of igniter landed in the grass next to the launch pad. A 15 mph wind was blowing. By the time the RCO saw the fire a few seconds after launch, it was out of control. The club then purchased a tarp for each launch area, and we never had another grass fire in the remaining six years I was with that club.

An additional precaution to take when a high fire danger exists is to bring several 5-gallon garden pump sprayers to the field. They should be filled with water and pressurized. Locate them in the range-head area. A club member is appointed as fire guard whose job it is to stand with the water sprayer just behind the modeler who's launching a model. The fire guard's eyes *never* look anywhere but at the launch pad, even when the model is launched. It is the job of the fire guard to *watch that launch pad* and to get out there with the water if by chance a conflagration starts.

(Be careful whom you assign as fire guard. Some people are exuberantly enthusiastic about the job, especially when it comes to completely soaking a competitor's launch pad and model on the slightest excuse.)

Do not set up rope barricades between the launch pads. Such barricades can be dangerous because they restrict the movement of people in the range-head area and give people a false sense of protection. On any model rocket range, you want to be able to move and *move fast* if something goes wrong. You don't want to find a barrier in the way. Flying on the range requires self-discipline. You do not cross another launch alley with its firing system wires and such. And when you hear a countdown in progress, either look up or get back. If the RSO and RCO don't have the good sense to call people on infractions of range-head discipline and safety or the guts to take disciplinary action when necessary, replace them.

Two portable folding tables are set up next to the main barricade gate as shown in Figure 18-2. At one of these tables is the range safety officer (RSO). Every model to be launched is presented to the RSO upon taking it into the range-head area. During contests, the RSO may delegate the checking duties to a safety check officer (SCO) and remain free to exercise continual safety vigilance over the range-head area.

During contests, the check-in table is also used for working up contest flight cards and for weighing the models. A battery-powered public address (PA) system should be available for use by the RCO. Years ago, we had to make our own PA systems. Now they can be purchased at places such as Radio Shack. Get a good 20-watt PA system capable of working from a 12-volt motorcycle battery. Modify the system so that the walkie-talkies used to communicate with trackers are plugged into the PA system. This gives *everyone* on the range the opportunity to hear the conversation and chatter with the trackers as well as giving the RCO a loud voice on the range. This method permits the range to be run by the RCO without requiring additional manpower for communicating with the trackers.

You should have at least *two* PA loudspeakers—four on a large range—mounted on a pole near the RCO table as shown. Point one toward the launch pads and the other toward the spectator area. Then everyone can hear what's going on. If you have a big range, mount the other loudspeakers on the left and point them also toward the pads and the spectators.

A range should have a flagpole—two if extra loudspeakers are used. The cheapest thing to use is a two-part TV antenna mast available at Radio Shack. If a TV antenna roof mount tripod is used, no guy ropes are required. Or you can make a portable flagpole from pipe and other common hardware items as shown in Figure 18-3.

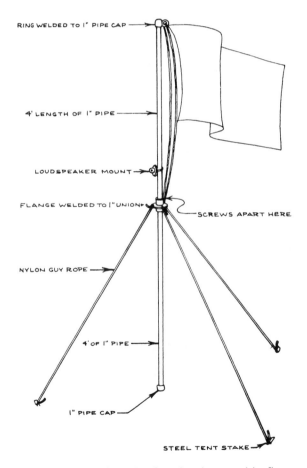

FIGURE 18-3   Sketch of a simple portable flag-pole and PA loudspeaker mount.

Some clubs have made flagpoles out of electrical conduit that can be quickly joined together to make 12-foot poles, then quickly disassembled when the range is knocked down after a day of flying. Two-foot lengths of pipe are buried flush with the ground (leave the bottom open so that rainwater doesn't accumulate in them) and the flagpoles are merely slipped into these during range setup.

Some ranges have several flagpoles because it looks so neat to fly Old Glory, the state flag, an NAR flag, the club flag, and the flag of any visiting club.

A flag is very important on a model rocket range. It draws attention to the fact that something is going on there. It locates the center of the action. It's also a wind indicator. And on the long trek back from recovering your model out in the boonies, it shows you where the range-head area is so that you don't get lost.

The equipment discussed thus far will cost about $200, depending on how much can be located in the junk piles of members' workshops, how much can be built by handy members, and how much can be obtained or scrounged from local donors. Be sure to design the range equipment so that it breaks down into small elements that are easy to carry and pack away for transportation and storage in foot-lockers, specially made wooden boxes, or other containers. Make your range equipment as portable as possible so that no more than two people are required to lift or carry any box. Design it so that it will fit into the back end of a station wagon.

By all means, build your range equipment so that it's *rugged*. It will take a beating. If you build it with this in mind, you won't have to spend a lot of time repairing and maintaining it. Good range equipment will last for ten years or more of hard use.

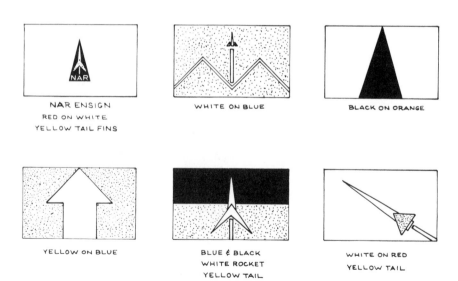

FIGURE 18-4   Some typical range flags of various NAR sections.

How does a model rocket range operate? The first order of business when everyone arrives on the range at the announced time—30 minutes prior to the start of flying time—is to set up the range. As few as six people can set up the complete model rocket range described here. However, the more people who are there, the faster the range will be set up. No one should be permitted to begin prepping models until the entire range is set up and checked out. If necessary, ensure participation by pulling setup assignments out of a hat if a call for volunteers doesn't elicit enough help. Every club member should be required to sign up for at least 30 minutes of range duties. A typical sign-up sheet is shown in Figure 18-5.

Although the range can be operated without trackers and with one person acting as both RSO and RCO if manpower is short, the complete range team usually consists of four people: RSO, RCO, and two trackers. The range safety officer (RSO) performs safety checking of models and controls the safety of the range as discussed earlier. On a busy range with lots of action, the RSO should be free to do nothing but be an RSO, and the safety check duties should be assigned to a deputy, the safety check officer, who checks for safety of models before they can be placed on the launch pads. The RSO should be at least 18 years of age. Why? Occasionally, the RSO is going to have to make a difficult safety decision that somebody else doesn't like. So

RANGE MANPOWER

Date: _____
Contest: _____

Time	Range Safety	Range Control	Track E	Track W	Return	Timers	Fire Guard	D	R
1:30–2:00									
2:00–2:30									
2:30–3:00									
3:00–3:30									
3:30–4:00									
4:00–4:30									
4:30–5:00									

FIGURE 18-5   Everyone who flies on the range should sign up on the Range Manpower Sheet for at least a half-hour shift during the flight session.

the RSO should be a person who can stick to a decision, often under considerable pressure from parents and other adults. Long, hard experience has shown that only a person over 18 should assume the position, duties, and responsibilities of an RSO.

The range-control officer (RCO) can be any club member. Although the job of the RSO is range safety, the RCO is in charge of operating the range itself. The RCO selects the modeler who will launch next, talks to the tracking stations, gives countdowns, records tracking data, and keeps a running commentary going on the PA system. In the crunch of a contest, a steady hand and a cool head are required of the RCO. However, it's a good idea for all club members to take a turn at RCO and learn how to do it, even those who are shy and quiet. Many people have learned how to handle themselves well in public and how to keep calm in a hassle because of their training and experience as an RCO.

Two trackers are required if tracking stations have been set up—or three trackers if the three-station elevation-angle-only system is used. If the rocketeers don't want their models tracked, don't set up and man the trackers. In that case, the range operates in the sport or duration mode. However, to give people experience and training in tracking, the trackers should be set up and manned for at least two 30-minute shifts during the flight session.

It's important to get all your club members qualified, experienced, and up-to-date on *all* range positions. Adults should be trained as RSOs, RCOs, and trackers. Young members should be trained as RCOs and trackers. New members should be urged to try a new range position each flight session, and older members should serve at least one 30-minute range duty shift during the flight session.

Once the range is set up and ready to go, everyone should sign up on the Range Manpower Sheet, such as that shown in Figure 18-5. *The range should not be opened for flying until all positions on all shifts are filled.* If you don't do this, you'll end up with a couple of people doing nothing but running the range all day. This isn't fair to them or to the other members who *should* be learning all the jobs.

Everyone who is going to fly picks a launch area and sets up their own launch pad, electrical firing system, and battery. This system stresses and teaches *individual responsibility.* Members are responsible for their own GSE. If it doesn't work, it's the member's fault, not the club's. If a modeler has ignition system or battery problems, you can bet that it will be fixed and working right by the time of the next flight session. It also stresses individual responsibility to follow range operating procedures.

Often two or more model rocketeers—especially members of a family—share the same launch area because each area has plenty of room to set up two launch pads if desired. Conflicts seldom arise because rarely do two modelers want to launch out of the same launch area at the same time. When this situation does occur and if it can't be resolved by the modelers themselves, the RCO can step in and assign priorities. (We discovered that four Stines required two launch areas when we were all flying in competition.)

There are two cardinal rules in the range-head area, and these are important for members to remember and for the RCO to enforce:

1. Never cut across the launch alleys. You could trip over somebody else's ignition wires.
2. Always approach the launch areas at right angles to the range-head area. Keep your head up and listen to the PA system. The RCO will be watching to ensure that you don't walk into a launch area that's in a terminal countdown, but don't depend on the RCO for your personal safety. Keep your eyes open and your head screwed on tightly.

Occasionally when a large model rocket is launched or a questionable one is about to go, the RSO and RCO will ask that adjacent launch areas be vacated during the terminal count. And they may call for a heads-up situation, so stop what you're doing and devote your full attention to the model being launched.

When you've prepped your model in the prep area and are ready to fly, bring your model through the gate to the range-head area and present it to the RSO or SCO for safety check. This is *always* done. If your model is to be tracked and you have a flight record sheet, bring it with you and present it to the RSO/SCO. The same holds true during a contest, when you should bring your contest flight card with you on the first flight and on subsequent flights pick it up at the return table.

The RSO, upon clearing your model for flight, permits you to proceed to your launch area with your model. Place your model on the launch pad, hook it up, and give it a final check. Then step back to your launch controller and raise your hand. On a busy range or with a sleepy RCO, you may have to holler, too. This signal alerts the RCO, who keeps a continual watch in the range-head area for people signaling that they're ready to fly by raising their hands. Normally, an RCO will work the launch areas in rotation, starting with number one and proceeding down the line, taking modelers as they're ready, and finally returning again to number one to sweep the range-head area again. This is the only fair way to operate when a lot of people want to fly because it's impossible for the RCO to keep track of who's next by the order in which they raise their hands.

At the very most, you'll have to wait only a few minutes until the RCO gets around to you. When your turn comes, the RCO gets verbal safety clearance from the RSO. Once the RSO has given this safety clearance, you may insert your safety key into your launch controller. The RCO then goes into the terminal countdown over the PA system. When the count reaches zero or start, push your ignition button and launch your model. It's then up to you to recover and return your own model.

If you have a misfire, *get your finger off the firing switch at once* and remove the safety key. Wait 1 minute before approaching the model on the launch pad. You don't have to take your model out of the range-head area to put a new igniter in it if you prepare ahead of time. Keep spare igniters in your pocket or at your launch controller. You can clear your misfire and, when ready, signal the RCO again that you're ready to launch.

If anyone launches without a safety clearance and countdown, the RSO declares that modeler out of action until the modeler finds the trouble and satisfies the RSO that everything is now okay. The RSO then lifts the no-fly restriction on the modeler.

Yes, I have seen occasional accidental launches because someone has a faulty ignition system. But because of the PA system, the RSO, the RCO, and the separation between launch pads, I've never seen one of these accidental launches become hazardous in thirty years. An accidental or premature launch is embarrassing, and the modeler quickly gets the equipment into safe shape. After all, the modeler can't blame anyone else!

Have I ever encountered any disciplinary problems on a rocket range? No. Everyone is there to fly and have fun. If people know and understand the rules, and if they see that everyone else is following the rules, they will follow them, too. I *have* seen modelers leave the range in a furious huff because they couldn't convince the RSO to clear a model for flight. But this is where the first rule of rocket ranges must always be observed: *The RSO is always right and doesn't have to answer to anyone on a safety decision.*

(One night at White Sands many years ago, I was range safety officer on an Aerobee sounding rocket flight. The winds aloft were so bad that they would blow the rocket off the range. I told the project scientist not to launch. He launched anyway. I pushed the button and blew up the rocket when it cleared the launcher. The next day, the scientist lodged an official complaint against me. My boss asked me what happened. I told him. Then my boss told me, "You did your job." That was the last I heard of it. But I must admit that I was a little apprehensive about it. An RSO often has to make tough decisions and stick by them.)

A tape recording of the flight operations on a range might sound something like this:

> *RCO:* Okay, the next bird to go is from launch area number 6. It's a single-stage model with a Type B motor and painted fluorescent orange with a white nose. Trackers?
> *Tracking East:* East go!
> *Tracking West:* West go!
> *RCO:* Range safety?
> *RSO:* Safety go!
> *RCO:* Range is go. Safety is go. Time is running at T-minus 5 . . . 4 . . . 3 . . . 2 . . . 1 . . . *start!*
> [The word *start* is a universal term used internationally to indicate the instant of ignition. It means the same thing in English, French, German, Dutch, Czech, Slovak, Polish, Romanian, Bulgarian, Italian, Spanish, Serbian, Greek, Hungarian, Russian, and Japanese. The word *fire* must *never* be used on a model rocket range unless there is indeed a fire in progress. At the call, "Fire!" coming from *any-*

where and *anyone,* everyone stops and helps put out the fire immediately.]

 RCO: Model coming up on peak. *Mark!* [The word *mark* indicates the moment of maximum altitude as seen by the RCO; both trackers should stop tracking at that moment whether or not it appears to them that the model has reached peak.] Recovery system is deployed. Descent looks good. Trackers, angles, please.

 Tracking East: [East always goes first.] Range Control, this is Tracking East. Azimuth, 3–2 degrees. [Azimuth angle is always reported first.]

 RCO: Roger. East, azimuth, 3–2 degrees. [The angles are always reported back to make sure the RCO has heard and recorded them correctly.]

 Tracking East: Elevation, 2–7 degrees.

 RCO: Roger. Elevation, 2–7 degrees. West?

 Tracking West: West, azimuth, 3–5 degrees.

 RCO: Roger, azimuth, 3–5 degrees.

 Tracking West: Elevation, 3–0 degrees.

 RCO: Say again Elevation. You were garbled.

 Tracking West: Elevation, 3–0 degrees.

 RCO: Still garbled. Elevation, 3–7 degrees?

 Tracking West: Negative, Range Control! Tracking West elevation is 3–0 degrees.

 RCO: Sorry. Got it that time. Elevation, 3–0 degrees. Okay, the next model to go is . . .

And so it goes. On a well-run range, as many as fifty models can be launched and tracked every hour from a 12-launcher range-head.

If, during the preflight or terminal countdown something goes wrong, *anybody* on the range may yell, "HOLD!" When the RSO and RCO hear this, they freeze the countdown right there, recycle the count back to the range safety clearance point, and find the reason for the hold call. This is an additional safety measure that makes *anyone* and *everyone* on the flying field a deputy range safety officer who can call a halt if something is seen to go wrong that the RSO or RCO may have missed. Naturally, this could get out of hand without good range discipline. Therefore, the RSO must *never* permit horseplay or false alarms on a model rocket range. It's the RSO's responsibility to maintain order and discipline. The RSO has the authority to have a person removed from the flying field for continued infractions of discipline.

In spite of model rocketry's outstanding safety record and all the safety procedures and checks, the nature of model rocketry means that some flights may not be successful. Because of good range safety discipline, I've never seen a serious hazard on a model rocket range. However, please keep in mind that you are dealing with objects that fly freely through the air at high speeds.

Therefore, I pass along to you the following tips from my own experience in watching about a half-million model rocket flights:

1. If you don't have range duties that require your attention, stop what you're doing during any terminal countdown and watch the model being launched. Keep your eyes on it until you know that the flight is going well and where the model is going to land.
2. Get on your feet and stay on your feet. Don't lie down and take a nap in the warm sun. You may have to move quickly to get out of the way of an errant model. It doesn't happen often. But when it does, be prepared.
3. If a model gets into trouble in the air, don't panic and run. Stand still and keep your eyes on it. If it comes in your direction, you'll have to move less than a foot to one side to get out of its way.

FIGURE 18-6   Finally, don't forget that you have to recover your own model rocket. A loyal and trustworthy recovery crew can be a model rocketeer's best friend.

4. Don't engage in horseplay or practical jokes on the range at any time, and don't let others do it, either.
5. Don't stand around in a crowd. During a heads-up flight, stand at least one arm's length away from everyone else.
6. If a model falls to the ground before its ejection charge has activated, stand clear of it until this occurs. Then, if it's yours, you may pick it up.
7. When recovering a model, don't run up on it. You may trip and step on it. This doesn't help the model very much. Don't recover somebody else's model unless you're asked to do so or unless you happen to find it out in the boonies while you're looking for your own.
8. Keep a clean model rocket range. Provide trash cans, and be tough on people who don't use them.
9. Help the RSO and RCO keep spectators under control. Don't let visitors wander into the range-head area or into the downrange recovery area. You and your group are responsible for the safety of everyone on the flying field. Your club will be a better one because of good range discipline, and spectators will respect your group because of it.
10. Keep dogs and small children under control at all times.
11. Don't try to fly on days with high winds and tinder-dry grass on the field.
12. Use common sense, keep your cool, and have fun.

I've been on many model rocket ranges all over the United States, Canada, and Europe. It's always a pleasure to be on a well-run range. I've walked off poorly run ranges. A smoothly running and efficient model rocket range where everyone takes turns at everything and everybody gets the chance to fly as much as time allows—well, *any* club or group can be proud of that.

19

CLUBS AND CONTESTS

There's a saying among model rocketeers that when two model rocketeers get together, they form a club. Actually, it takes more than two people to have a model rocket club. A club has a lot of advantages. If you don't belong to one, you should seriously consider forming one or joining one.

It's easiest to join a club that's already in existence. But if there isn't one around, it isn't difficult to form your own. In fact, your club may come into being spontaneously when several people meet on the same field to fly together. They begin to enjoy one another's company. And they decide to expand the group to include others who enjoy model rocketry, too.

How do you get a model rocket club going? How do you keep it going? How do you make it a worthwhile activity for its members—one that they'll support by coming to meetings and flight sessions and by giving their time and effort?

Let's suppose you're a lone rocketeer who's done some flying on your own, observing the proper safety rules. You'd like to learn more about the hobby, discover what other people are doing, and encourage others to do it the right way. If you're a young person, the first thing you should do is to find an adult who's also interested in forming and belonging to a model rocket club. The adult can serve as the club adviser. As I pointed out in the preceding chapter, an adult can do many things that a young person cannot. Your adult adviser may turn out to be one of your parents, a science teacher, a Scout or 4-H leader, or somebody at the hobby shop who's interested in model rocketry.

You should also try to find a sponsoring organization. A sponsor can help you find a meeting place, a flying site, and the seed money or financial assistance to get the club started. Many organizations today have model rocketry programs within their own structure or warmly encourage model rocket club activity. The Civil Air Patrol has an active model rocket program. Many Scout and explorer groups are deeply involved in model rocketry. The 4-H Clubs also have a model rocket program. But many other organizations are potential candidates. These include but are not limited to Lions, Kiwanis, Rotary International, the local fire department, the local police department, and the YMCA.

It will be of immense help to you in organizing the club, obtaining a sponsor, and finding meeting and flying sites if you, your senior adviser, and all your club members are also members of the National Association of Rocketry (NAR). This is a nationwide organization of model rocketeers. You can write to the NAR at the headquarters address listed in Appendix I. The organization will send you membership information, membership application blanks, and information on how to get an NAR charter for your club.

I'm proud to be one of the founders of the NAR in 1957. It's grown from an idea to an organization with national and international respect. Today, it has thousands of members and chartered NAR sections all over the United States.

The NAR does many jobs. It's a nonprofit organization. This means it's a labor of love for the people who run it. The most important NAR offering for club purposes is liability insurance to protect you, your sponsor, and the flying site's landowner. It will cover personal injury and property damage caused by a model rocket if an accident should happen. Having insurance doesn't mean that model rocketry is dangerous. Quite the contrary. If it were dangerous, no one would offer to insure model rocketeers at all!

FIGURE 19-1   The National Association of Rocketry (NAR) offers the model rocketeer many membership benefits. Many clubs are chartered NAR sections. Here are some NAR publications. Jacket patches come from various clubs and contests.

Having this liability insurance will be a tremendous help in obtaining the support of sponsoring organizations and getting permission to use meeting rooms and flying fields. For a few extra dollars, the insurance coverage can be extended to third-party coverage to protect the sponsor and the flying site owner. Believe me, NAR insurance is a big factor when you're looking for support. Full details are available from NAR headquarters.

The NAR also charters local clubs as official sections of the NAR. It's not easy to get an NAR section charter. Stiff requirements must be met by the club. It must have a minimum number of members, one of whom must be an adult over age 18. The club must petition the NAR for a section charter, listing the names and NAR membership sporting license numbers of the club members. It must submit a set of operating rules called bylaws that must be checked and approved by the NAR. The activities of the group must be reported to the NAR at regular intervals.

The NAR wants its sections to be active, operating groups of serious model rocketeers. NAR sections are not flash-in-the-pan neighborhood rocket clubs that have sprung up overnight and are likely to fold as quickly as they were formed. Sections must have organization, advisers, proper direction, and the ability to last. None of this is impossible to accomplish.

Bylaws should be adopted at your earliest meeting. They are the organizational and procedural rules of the club, a document that club members turn to for guidance. A sample set of recommended NAR section bylaws is presented in Appendix VIII. Some club bylaws are simpler than this, some more complex. But any set of bylaws should contain the provisions given in the sample. Note that the bylaws don't contain any special rules for the operation of the club's model rocket range or for the construction and flying of models by members. The basic NAR Model Rocket Safety Code, rules, regulations, and standards cover these operational aspects. Special club standards have been adopted by some groups, but these should be separate documents.

Note that the flight sessions aren't considered to be meetings of the club. It's impossible to conduct a business meeting on a model rocket range. Everyone is too busy with models and range duties. Business meetings of the club are important and should be held regularly in a place where people can be comfortable and discussions can be held without distractions.

Although some financial support may come from sponsoring organizations from time to time, a club should try to become financially solvent. It should have its own membership dues to cover the costs of mailing meeting notices and for other general club operating expenses. Most clubs charge nominal dues of a couple of dollars per month. These dues should be over and above those to other organizations such as the NAR.

One of the most important aspects of club operations is good communications between officers and members. There should be a means for rapidly getting information to every member of the club. This may be done very easily by e-mail. Dates for meetings and flight sessions should be regularly scheduled and announced well in advance. If there's a change in plans, such as cancellation of a flight session because of weather, the telephone committee goes into action. A monthly business meeting is a must. Flight sessions should be held once a month.

A club newsletter shouldn't be a substitute for face-to-face communications at a club meeting. A newsletter is more a means of publishing a record of meeting and flying schedules, upcoming contests, and other club data that should exist as hard copy. Also, the club newsletter should be a means of letting *other people* know what you're doing. It should be sent to all members as well as to local newspapers, radio and TV stations, public officials, and officers of sponsoring organizations.

Business meetings should be short and to the point without getting tangled up with personality clashes, ego trips, and the puzzlements of *Robert's Rules of Order.* Elect a president or chairperson who can keep the meeting going and doesn't like to give speeches. Get in touch with NASA to obtain videos and other attractions to make your meetings sparkle with a program that's of interest to model rocketeers. If you do this, you'll have very little trouble getting your members to come to meetings.

Over the years, I've formed or been a member of six model rocket clubs. In all cases, we've always faced a serious problem: teaching the newcomers. So in 1965, I embarked on an experiment that ran for eight years and turned out to be highly successful. I formed the New Canaan (Connecticut) YMCA Space Pioneers Section of the NAR. Every new member had to meet a series of tough qualifications. People under age 12 had to have a parent or adult join with them and participate with them in all club activities. Attendance at meetings and flight sessions was mandatory unless the member was excused in advance. Membership was open only once each year in September. The reason for this was the requirement that every club member complete the club's nine-month training course in model rocketry taught by experienced club members. We followed a regular syllabus or course plan and used earlier editions of this book as the text. The training course was intended to "bring everyone up to the same level of ignorance and confusion that the other club members enjoyed." As you might guess, we had fun. The training course provided every member with the same foundation upon which to build future activities in model rocketry. It eliminated a lot of trial-and-error activity by new members. It did *not* squelch creativity or individual choice. Out of this program came many developments that are taken for granted today in model rocketry: the present range operating system, simple beginners' models that became commercial kits, simple contest events, simplified methods of calculating CP and estimating altitude, and a number of other things that are commonplace in the hobby now.

The training course made use of existing kits available in local hobby stores. Every two weeks, we held a 45-minute lecture covering an assigned subject or chapter of this book, often with demonstrations. A flight session was held two weeks later to let the trainees try out the things they'd studied and built. There were regular assignments of outside work involving reading in this book and building models in the members' own workshops. There were never any workshop sessions where trainees brought their models to the club meeting to work on them together; our meeting time was far too valuable and was used instead to present information to the trainees. Once the lecture session was over, the older, trained members arrived for business discussions. The flight sessions were the testing periods during which the trainees proved how well they were getting along. Organized sporting competition was used to keep the work interesting and *to keep the experienced members on their toes.* Often, the trainees were pitted against the older members in contests—and won! Awards, trophies, medals, certificates, and merchandise prizes were given. Upon completion of the course, each trainee was given a humorous diploma certifying him or her as a "Compleat Model Rocketeere."

Members who had gone through the training course went on to do their own thing—competition, research, or advanced sport modeling. How successful was the training course concept of club operations? During every year of the club's existence between 1965 when it was formed and 1973 when I left to move to Phoenix, there was at least one U.S. champion model rocketeer in the club. Twice, the New Canaan YMCA Space Pioneers won the coveted NAR national championship section pennant. One club member became an international medalist in the First World Championships for Space Models in Yugoslavia in 1972.

I've also used the training course system in two other clubs. It works. This statement is backed up by facts. Sometimes, the procedure must be modified because of local conditions. For example, in Phoenix, we could fly all winter and had to cut back on flying sessions during the summertime with 110 degree F temperatures.

Many people have criticized the strong discipline and highly structured operation of the training course approach. In today's more permissive environment, it doesn't fly as well. It would have to be modified. However, I've seen what it can accomplish with young people. *It's what a model rocket club is really good at doing!*

Out of the training course concept experience, I developed the NAR's NARTREK (NAR Training of Rocketeers for Experience and Knowledge) program that consists of a series of different achievement levels—bronze, silver, gold, and advanced—with standards to be met at each level. I highly recommend the NARTREK program for clubs because it solves the problem of restricting new members to admission only once a year at the start of the organized training course. Using NARTREK, each participant proceeds at his or her own pace. This book can be of great help in proceeding through NARTREK because the program itself is based on this book, which in turn is based on what I learned in the New Canaan YMCA Space Pioneers training program. Your club meetings can be used to answer questions or to cover specific aspects of the NARTREK program with which members may be having problems. When all club members have won the coveted NARTREK gold badge, you've got an experienced club! Ask NAR headquarters about the NARTREK program.

Your club will probably be called upon to give flight demonstrations (demos) from time to time. These may be held simply to amaze and impress

FIGURE 19-2 Model rocketeers of all ages travel across the United States every year to the National Model Rocket Championships (NARAM) or to national sport flying fests. These are fun affairs full of camaraderie where people help one another and make new friends, young and old.

your friends, schoolteachers, or the public. Or they may have a more important function such as getting new members or showing local public safety officials what model rocketry is all about. Most model rocket manufacturers will be glad to help you out on a big demo, and some have special demo support programs. Write to them about this. But do it a month or more in advance, please.

When you put on a flight demo, do so without an admission charge, even though your club treasury could stand some bolstering. Pass the hat for donations instead. Then you'll eliminate all sort of hassles about entertainment taxes and such.

A demo is the time to show off your club shirts or hats, if you have them. It's a good idea to have club shirts or hats because it will allow you to tell a member from a spectator. Such confusion has happened. Club shirts or hats also identify club members to the RCO when they're in the downrange area. And they tell spectators who's a member in case they have any questions.

It's important to maintain strict range discipline during a demo. It's also important to fly only models of very high reliability. A demo is no place to try out an experimental design for the first time. Safety checking should be extra strict.

A short, successful demo is best. Start with two low-altitude models: one with a streamer and the other with a parachute. Follow this with a cloud-buster, a high-performance little model powered by a Type C motor. Fly a B/G, a two-stage model, and perhaps a large model rocket if the demo field or area is big enough. Payload models are always impressive and show that model rocketry is more than up-and-down. Fly an egg, squeezing the flight for all the drama and suspense possible. Save your best for the last, leaving the spectators with something amazing to talk about. *Don't* fly salvos where two or more models are launched simultaneously; that's fireworks-type stuff. Have some-one talking on the PA system at all times, telling people over and over again what's going on. Use the full countdown ritual and take every opportunity to show your spectators that model rocketry isn't a bunch of kids playing with toys but a serious technical hobby, a technology in miniature, the Space Age brought to Main Street USA.

Remember: more people have seen model rockets launched than have ever watched a live launch in person from the Cape. What they see at your demos and flight sessions is, to them, the space program in miniature. Your model rockets may be the *only* rocket vehicles they've ever seen!

Contests provide lots of fun in model rocketry. Competition began when one model rocketeer said to another, "My model rocket will go higher than your model rocket!" And the other rocketeer retorted, "Prove it!"

To keep your club's first contest from becoming too hectic and confused, schedule only two simple events such as spot landing and parachute duration.

FIGURE 19-3   One thing you are certain to always see at any club launch is *lots of rockets*!

Very little special equipment is required for these events—a 100-foot tape measure and a brightly colored pole for spot landing, and two stopwatches. for the duration event. You should have three judges who are adults. They will have to render impartial decisions on all the little protests and complaints that will arise. Such minor hassles are always part of competition, even with the best aura of sportsmanship. The rules should be published or given to each competitor in advance, and you should hold a precontest briefing on the range to make sure everybody understands the rules and operating procedures *before* the contest starts. Then, don't change the rules or procedures during the contest because this could change the nature of the contest and put some people at a disadvantage. If weather causes the contest to be stopped in the middle, either close the contest and award prizes on the basis of what's happened thus far, or reschedule the contest and start the whole thing again from square one at a later date. Otherwise, different weather, wind, field, or operational conditions could give competitors who have flown an advantage or disadvantage over those who didn't get a chance to fly.

The rules for flight duration events can be simple. Each flight should be timed by two timers with stopwatches in case something happens to one stopwatch. Timers start their stopwatches when the model takes off and stop the watches when the model goes out of sight, touches the ground, or lands in the branches of a tree. Typical duration events include streamer duration, parachute duration, and egg loft duration. You should place a limit on the total impulse that may be used but no limit on streamer or parachute size, wing area, etc. This will encourage people to try different approaches to winning. You can even conduct simple altitude events using the marker streamer method as explained in chapter 17.

Spot landing events are fun. There should be no limit on total impulse. Only one flight should be allowed, although the contestant can practice as much as desirable. But when the modeler says, "This is my official flight," that's it. Place the spot landing pole at a place on the field selected at random, usually a site not closer than 100 feet to the range-head area and usually in the downrange recovery area. Tape or staple a colored streamer to the top of the pole as a wind indicator. Most important rule: *All* recovery devices must deploy fully before the model touches the ground because the pole is *not* a target. This is a spot landing contest and the pole is the spot. And don't let modelers tilt their launch rods more than the permissible 30 degrees from the vertical; use a protractor with a weighted string to check launcher tilt if any questions arise. The winner is the one who lands the model with the tip of the nose cone closest to the pole.

If you want to run your contest under the standardized national rules, the NAR publishes the U.S. Model Rocket Sporting Code. All NAR-sanctioned contests must be conducted according to these rules. If your club is an NAR section, you'll get a lot of information about running a contest from the NAR Contest and Records Committee. Having all contests run in the same fashion with the same rules on a nationwide basis makes it possible for your club to fly in somebody else's contest if it's an open or regional meet (section meets are

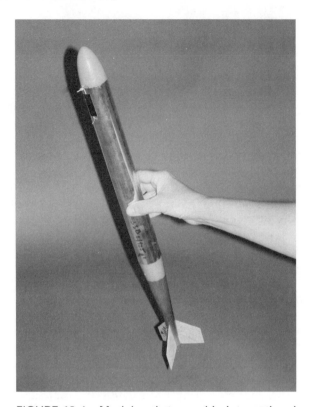

FIGURE 19-4   Model rockets used in international competition are very technically advanced. This 40-mm S3B parachute duration model is constructed from hand-layed-up fiberglass and epoxy. The model is wrapped in a silver reflective Mylar to improve visibility and colored tissue to reduce porosity of the fiberglass weave. It uses a micro-dethermalizer (DT) to bring the model down just after the 7-minute maximum flight time. It was flown in the 14th World Spacemodeling Championships in the Czeck Republic by U.S. team member Mark Petrovich.

those sanctioned contests open only to club or section members). It won't take long for your club to be ready to take on another club in a sanctioned contest on another field. This can be a fun weekend for everyone. And some great club rivalries have sprung up over the years.

Then there's the big annual national meet, NARAM (NAR Annual Meet), where hundreds of model rocketeers from all over the United States congregate. This is the oldest continuously held national model rocket competition in the world. The first, NARAM-1, was held in Denver, Colorado, in 1959. Since then, NARAMs have been held at the U.S. Air Force Academy, NASA

Wallops Flight Center, the U.S. Army Aberdeen Proving Ground, the NASA Johnson Space Center, and many other Space Age sites.

If you're not into competition, then the sport launches may be your fun thing. People just get together to fly for fun. Often, an unofficial contest such as sport scale or egg lofting duration is flown on the side for fun with no contest points but with ribbons and merchandise awards given instead. Regional and national sport meets are becoming more common.

The ultimate challenge in competition is to earn the right to be called world champion. Internationally, model rocketry is called spacemodeling. Spacemodeling championships have been held every two years since 1972. The Federation Aeronautique International (FAI) is the governing world body, covering not only space modeling but also all aspects of aviation. Each member country is represented within the FAI by its national aero club, which in the United States is the National Aeronautic Association (NAA). The NAA delegates authority for space modeling to the Academy of Model Aeronautics (AMA). The AMA sponsors the U.S. Spacemodeling Team.

Spacemodeling world championships are regularly attended by over twenty-five nations with several hundred competitors vying for gold medals. There are two age levels of competition: juniors (17 years or younger) and seniors (18 years or older). The competition is intense but is often overshadowed by the unique cultural exchange between teams. At the end of each championship, there is always an awards banquet with plenty of great storytelling and camaraderie. To learn more about the FAI World Spacemodeling Championships, go to www.spacemodeling.org.

In clubs, contests, and sport flies lie the fun and advancement of the sport of model rocketry. Clubs give you the opportunity to learn how to work with people, something that in the long run may be far more important than knowing about rockets and other technical hardware. Contests run under accepted rules separate the good modelers from the not-so-good modelers and reward them accordingly. Competition improves things. And sport flying sessions get you together with other people who have the same interest in and, yes, love of model rocketry.

20

WHERE DO I GO FROM HERE?

Model rocketry presents many technical challenges—in fact, more than you could ever master or try during your lifetime as a model rocketeer. But there has always been an underlying urge to build rockets that are bigger and fly higher or faster than the legal limits of a model rocket (remember, no more than 3.3 pounds [1,500 grams] at liftoff and no more than 4.4 ounces [125 grams] of propellant). Beyond model rocketry, there is high-power rocketry (HPR). High-power rockets can use no more than 40,960 newton-seconds (9,204 pound-seconds) of total impulse in the rocket motor and must be flown in compliance with Federal Aviation regulations. These are *big rockets!* I've seen rockets that weighed more than 25 pounds and flew over 5,000 feet.

High-power rocketry is a safe activity with a proven track record. It has its own safety code, and a model code for its use has been developed by the National Fire Protection Association (NFPA 1127 Code for High-Power Rocketry).

High-Power Rocketry Safety Code
1. *Certification.* I will fly high-power rockets only when certified to do so by the National Association of Rocketry.
2. *Operating Clearances.* I will fly high-power rockets only in compliance with Federal Aviation Regulations Part 101 (Section 307, 72 Statute 749, 49 United States Code 1348, "Airspace Control and Facilities," Federal Aviation Act of 1958) and all other federal, state, and local laws, rules, regulations, statutes, and ordinances.
3. *Materials.* My high-power rocket will be made of lightweight materials, such as paper, wood, rubber, and plastic, or the minimum amount of ductile metal suitable for the power used and the performance of my rocket.

4. *Motors.* I will use only commercially made, NAR-certified rocket motors in the manner recommended by the manufacturer. I will not alter the rocket motor, its parts, or its ingredients in any way.

5. *Recovery.* I will always use a recovery system in my high-power rocket that will return it safely to the ground so that it may be flown again. I will use only flame-resistant recovery wadding if wadding is required by the design of my rocket.

6. *Weight and Power Limits.* My rocket will weigh no more than the motor manufacturer's recommended maximum liftoff weight for the motors used, or I will use motors recommended by the manufacturer of the rocket kit. My high-power rocket will be propelled by rocket motors that produce no more than 40,960 newton-seconds (9,204 pound-seconds) of total impulse.

7. *Stability.* I will check the stability of my high-power rocket before its first flight, except when launching a rocket of already proven stability.

8. *Payloads.* My high-power rocket will never carry live animals (except insects) or a payload that is intended to be flammable, explosive, or harmful.

9. *Launch Site.* I will launch my high-power rocket outdoors in a cleared area, free of tall trees, power lines, buildings, and dry brush and grass. My launcher will be located at least 1,500 feet from any occupied building. My launch site will have minimum dimensions at least as great as those in the launch site dimension table (Table 10). As an alternative, the site's minimum dimension will be one-half the maximum altitude of any rocket being flown, or 1,500 feet, whichever is greater. My launcher will be no closer to the edge of the launch site than one-half of the minimum required launch site dimension.

10. *Launcher.* I will launch my high-power rocket from a stable launch device that provides rigid guidance until the rocket has reached a speed adequate to ensure a safe flight path. To prevent accidental eye injury, I will always place the launcher so that the end of the rod is above eye

TABLE 10
HPR Launch Site Dimensions

Total Impulse All Engines (newton-seconds)	Equivalent Motor Type	Minimum Site Dimensions (ft.)	Equivalent Dimensions (miles)
160.01–320.00	H	1,500	0.25
320.01–640.00	I	2,500	0.5
640.01–1,280.00	J	5,280	1
1,280.01–2,560.00	K	5,280	1
2,560.01–5,120.00	L	10,560	2
5,120.01–10,240.00	M	15,840	3
10,240.01–20,480.00	N	21,120	4
20,480.01–40,960.00	O	26,400	5

level, or I will cap the end of the rod when approaching it. I will cap or disassemble my launch rod when not in use, and I will never store it in an upright position. My launcher will have a jet deflector device to prevent the motor exhaust from hitting the ground directly. I will always clear the area for a radius of 10 feet around my launch device of brown grass, dry weeds, or other easy-to-burn materials.

11. *Ignition System.* The system I use to launch my high-power rocket will be remotely controlled and electrically operated. It will contain a launching switch that will return to "off" when released. The system will contain a removable safety interlock in series with the launch switch. All persons will remain at a distance from the high-power rocket and launcher as determined by the total impulse of the installed rocket motor(s) according to the safe distance table (Table 11).

12. *Launch Safety.* I will ensure that the people in the launch area are aware of the pending high-power rocket launch and can see the rocket's liftoff before I begin my audible 5-second countdown. I will use only electrical igniters recommended by the motor manufacturer that will ignite rocket motors within 1 second of actuation of the launching switch. If my high-power rocket suffers a misfire, I will not allow anyone to approach it or the launcher until I have made certain that the safety interlock has been removed or that the battery has been disconnected from the ignition system. I will wait 1 minute after a misfire before allowing anyone to approach the launcher.

13. *Flying Conditions.* I will launch my high-power rocket only when the wind is no more than 20 miles per hour and under conditions where the rocket will not fly into clouds or when a flight might be hazardous to people, property, or flying aircraft. Prior to launch, I will verify that no aircraft appear to have flight paths over the launch site.

14. *Prelaunch Test.* When conducting research activities with unproven designs or methods, I will, when possible, determine the reliability of my high-power rocket by prelaunch tests. I will conduct the launch-

TABLE 11
HPR Safe Distances

Total Impulse All Engines (newton-seconds)	Equivalent Motor Type	Minimum Distance from Rocket with Single Motor (ft.)	Minimum Distance from Rocket with Multiple Motors (ft.)
160.01–320.00	H	100	200
320.01–640.00	I	100	200
640.01–1,280.00	J	100	200
1,280.01–2,560.00	K	200	300
2,560.01–5,120.00	L	300	500
5,120.01–10,240.00	M	500	1,000
10,240.01–20,480.00	N	1,000	1,500
20,480.01–40,960.00	O	1,500	2,000

ing of an unproven design in complete isolation from persons not participating in the actual launching.

15. *Launch Angle.* I will not launch my high-power rocket so that its flight path will carry it against a target. My launch device will be pointed within 20 degrees of vertical. I will never use rocket motors to propel any device horizontally.

16. *Recovery Hazards.* If a high-power rocket becomes entangled in a power line or other dangerous place, I will not attempt to retrieve it. I will not attempt to catch my high-power rocket as it approaches the ground.

High-power rocketry is a regulated activity. You must be at least 18 years old to purchase HPR rocket motors (as mandated by the U.S. Consumer Product Safety Commission [CPSC]). Additionally, the sale and storage of HPR rocket motors is regulated by the Bureau of Alcohol, Tobacco and Firearms (BATF). Manufacturers and dealers are not allowed to sell HPR motors to a person who is not certified by the National Association of Rocketry (NAR) or the Tripoli Rocketry Association (more on certifications later). An airspace waiver must be obtained from the Federal Aviation Administration (FAA) for all HPR flights. In conjunction with these regulations, HPR rockets are normally flown at scheduled launches throughout the United States. This eases the burden of highly restricted and costly shipment of the rocket motors (dealers deliver them to the launch for customers), approved magazine storage requirements (BATF has regulations limiting storage of HPR motors in your home), and allows large numbers of rocketeers to take advantage of FAA waivers for a specific site and time period. Besides, flying HPR rockets with a group of other rocketeers is fun! Who wants to spend upward of several hundred dollars on an HPR rocket and motor and not have any fellow rocketeers with whom to witness the flight? Some of the largest annually scheduled events draw a thousand people.

HPR rockets basically are scaled-up model rockets with very strong construction. A minimum-diameter rocket with the correct motor can achieve speeds in excess of the speed of sound! Altitudes of over 30,000 feet have been verified. But most HPR rockets are simply larger airframes with larger total-impulse motors that fly to average altitudes of 5,000 feet (that's still almost a mile!)

The best way to get started in high-power rocketry is to attend an HPR launch and ask lots of questions (www.rocketryonline posts a launch calendar). Then you should start with a commercially available kit. There are many manufacturers of quality kits, including Public Missiles Ltd. and LOC Precision. They both offer a full line of sturdy, well-designed basic kits. You'll find many of the materials used in HPR rocket kits are also used in real sounding rockets flown by NASA: G10 fiberglass, carbon fiber, and Kevlar, to name a few.

Recovery of HPR rockets is an added challenge. The flying site minimum dimensions are much greater than for model rockets. Most typical HPR flying sites are a square mile. Remember learning about rocket-eating trees earlier in this book? The number of good HPR flying sites increases as you head west

FIGURE 20-1   Electronics are commonly used to deploy recovery systems in high-power rocketry. This altimeter is a Missileworks RRC2 mounted on an electronics bay board cut by Precision G10.

across the country. One of the cleverest uses of technology to solve such a simple problem as not chasing your HPR rocket that just flew to 5,000 feet for miles downrange utilizes a small onboard computer to control the deployment of a small drogue parachute at or just after apogee. See Figure 20-1. Then the rocket descends quickly but safely to a predetermined lower altitude (usually 300 to 400 feet AGL), and the main parachute is deployed. The rocket often comes within a few hundred feet of where it was launched!

Just like model rocketry, high-power rocketry is based on the principle that you never make your own rocket motor. Leave the design and manufacturing hazards to the professionals. There is a huge variety of commercially available HPR motors. Typical motor diameters used in HPR are 29 mm, 38 mm, 54 mm, 75 mm, and 98 mm. Motors can be single-use or reloadable. Single-use motors are factory assembled and have accurate total impulse and delay that cannot be altered. A reloadable motor has much more flexibility by the user in impulse and delay selection and is generally much cheaper to use on a flight-by-flight average.

One of the most exciting and innovative motor technologies has been developed for HPR: hybrid rocket motors. A hybrid rocket motor uses a solid fuel and liquid oxidizer. The fuel is often a simple, thick cardboard or plastic tube, and nitrous oxide (NOX) is the most common oxidizer. High-power rocketeers love the added complexity of a hybrid motor system. The up-front cost of the necessary ground support equipment and hybrid motor itself can be several hundred dollars, but the cost per flight is very low. To learn more about hybrid rocket motors, go to www.flyhybrids.org.

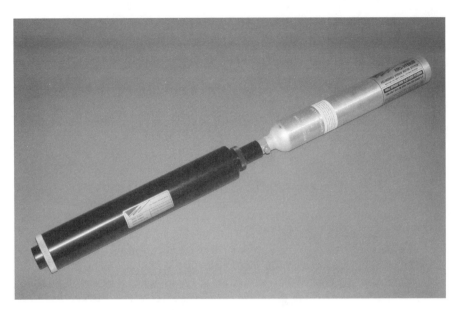

FIGURE 20-2   The Aerotech hybrid motor system where the oxidizer tank is attached to a conventional reloadable motor casing. The tank is loaded before it is attached and installed in the rocket. A pyrovalve releases the oxidizer into the combustion chamber when the igniter is fired.

Great! Now you're excited and want to build your first high-power rocket. Before you go any further, you'll need to get your user certification from the National Association of Rocketry or the Tripoli Rocketry Association. There are three levels of NAR high-power user certification:

Level 1—for H and I motors

Level 2—for J, K, and L motors

Level 3—for M, N, and O motors

Each level must be completed in succession. After you become Level 1 certified, you are under no obligation to advance to higher levels. It is up to you.

NFPA 1127 allows an uncertified individual to purchase a single motor for certification purposes, but make sure your motor supplier knows you're buying the motor for a certification attempt. This is the only condition under which uncertified individuals can obtain high-power rocket motors.

Level 1 Certification

1. Complete an NAR high-power user certification form. You can get the form in PDF format at the NAR Web site (www.nar.org/cabinet/hpappl.pdf) or by contacting NAR headquarters and requesting a user certification packet.

2. Assemble your certification team to witness your flight. The team must consist of either two NAR members, one of whom must be Level 1 certified, or a single NAR member who is Level 2 certified. Most NAR sections have certified members who can help you.
3. Make a safe and successful flight of a rocket using either an H or I motor.
4. Have your certification team sign your forms, and return them to NAR headquarters.
5. The forms will have a temporary HPR certification license so that you can continue flying at Level 1. Within a few weeks, you'll receive a new NAR membership license with your Level 1 certification printed on it.

Level 2 Certification

1. Complete an NAR high-power user certification form.
2. For Level 2 certification, you must successfully complete a brief written exam. You will be tested on your knowledge of the regulations and laws pertaining to high-power rocketry and rocket technical knowledge—for example, center of pressure and centure of gravity relationships.
3. The examination consists of thirty-three questions in multiple-choice format. The passing grade is 88 percent. The test may be taken only once in a thirty-day period and must be completed prior to the flight requirement.
4. The entire pool of test questions is available to you in an online interactive test. Visit the NAR Web site at www.nar.org/hpcert/hptest_part_A.html to begin your review of the questions.
5. Assemble your certification team to witness your flight. The team must consist of two Level 2–certified NAR members or one Level 3–certified NAR member.
6. Make a safe and successful flight of a rocket using either a J, K, or L motor.
7. Have your certification team sign your forms, and return the signed documents to NAR headquarters.
8. The forms will have a temporary HPR certification license so that you can continue flying at Level 2. Within a few weeks, you'll receive a new NAR membership license with your Level 2 certification printed on it.

Level 3 Certification

Level 3 certification qualifies you to purchase and use M, N, and O motors. Qualification criteria are much more stringent than for Levels 1 and 2. You can review the full procedure by obtaining a PDF document (www.nar.org/cabinet/l3pkg.pdf) at the NAR Web site or by contacting a member of the Level 3 certification committee. A brief summary of the Level 3 requirements is as follows:

1. You must be an NAR Level 2–certified flyer.
2. Your rocket must be substantially built by you.
3. The rocket must contain a redundant mechanism other than the motor's ejection charge for performing the initial recovery event.
4. The capability must exist to externally disarm all pyrotechnic devices in the rocket.
5. The flyer must obtain and fill out an NAR Level 3 certification form.
6. This form documents the certification procedure steps. The flyer must also prepare a certification package as defined in these requirements.
7. During the construction of the rocket, the flyer must complete the rocket construction packet for review by the Level 3 certification committee.
8. Prior to the certification flight, the flyer must present a recovery systems package to one of the Level 3 certification committee members.

FIGURE 20-3   This 3X scaled-up Sumo was built by Dirk and Eric Gates. The rocket is powered by a total of four M motors. Inset photo: Eric is installing a digital video camera prior to launch at LDRS-21. (Photo by Gary Rosenfield.)

FIGURE 20-4 David Schaefer built this 10-foot-tall radio-control, glide-recovery X-30. Several smaller test models were built to test the design and flight characteristics.

9. At the time of the certification flight, the flyer will present a completed certification package for approval as described in the certification package guidelines.
10. Upon approval of the certification package, the flyer must make a successful certification flight as described in the certification flight requirements.
11. Upon successful completion of the certification flight, the completed approved certification package must be sent to NAR headquarters for final processing as described in the final procedures after certification.
12. The forms will have a temporary HPR certification license so that you can continue flying at Level 3. Within a few weeks, you'll receive a new NAR membership license with your Level 3 certification printed on it.

High-power rocketry is as close to real sounding rockets as you can get (if your desire and wallet can afford it). The safety protocols and certification procedures are designed to increase your safety and enjoyment. *Always* follow them just like professional rocket engineers do!

EPILOGUE

By now, it should be apparent that there's a great deal to this technological recreation called model rocketry. As many people have discovered, it involves nearly every aspect of human endeavor, just like its full-size counterpart in astronautics. I've only touched the surface of model rocketry in this book. It's up to you to go on from here.

As you progress in model rocketry—or anything else you do during your life—teach someone else what you've learned. Communicate. Write it down. If I've helped you in this book, and if others have helped you, the only way you can repay us is to do the same for somebody else. Don't pay back; pay forward. That's been the basis for the rise of civilization for a hundred centuries.

How far do you want to go in model rocketry? That's up to you. If you have read, understood, and followed all that I've tried to tell you about here, you're on your way!

BIBLIOGRAPHY

Abbott, Ira H., and Von Doenhoff, Albert E. *Theory of Wing Sections*. New York: Dover Publications, 1959.

Alway, Peter. *Rockets of the World: A Modeler's Guide*. P.O. Box 3709, Ann Arbor, Mich. 48106, 1993.

American Radio Relay League, Inc. *The Radio Amateur's Handbook*. West Hartford, Conn., current edition (revised annually).

Banks, Michael A. *Advanced Model Rocketry*. Milwaukee, Wisc.: Kalmbach Publishing Company, 1985.

Banks, Michael A., and Cannon, Robert L. *The Rocket Book*. Milwaukee, Wisc.: Kalmbach Publishing Company, 1985.

Canepa, Mark B. *Modern High Power Rocketry*. Victoria, British Columbia, Canada: Trafford Publishing, 2002.

Gregorek, Gerald M. *Aerodynamic Drag of Model Rockets*, Estes Model Rocket Technical Report TR-11, product number EST 2843. Penrose, Col.: Estes Industries, Inc.

Hoerner, Sighard F. *Fluid Dynamic Drag*. 148 Busteeds Drive, Midland Park, N.J.: Dr. Sighard F. Hoerner, 1958.

Hurt, H. H., Jr. *Aerodynamics for Naval Aviators*, NavAir 00-80T-80, Washington, D.C.: Government Printing Office, 1965.

Mandell, Gordon K. *et al. Topics in Advanced Model Rocketry*. Cambridge: MIT Press, 1973. (Reprinted by and available from the NAR.)

Pratt, Douglas R. *Basics of Model Rocketry*. Milwaukee, Wisc.: Kalmbach Publishing Company, 1993.

VanMilligan, Timothy S. *Model Rocket Design and Construction*. Colorado Springs: Apogee Components, Inc., 2000.

Zaic, Frank. *Circular Airflow and Model Aircraft*. Northridge, Calif.: Model Aeronautical Publications, 1964.

Note: The National Association of Rocketry (NAR) has a large and growing number of technical reports and other publications available to its members through the NAR Technical Services (NARTS). Estes Industries, Inc., and Quest Aerospace, Inc., have a large and growing number of publications and software products available. The amount of information in model rocketry is probably greater than in any other hobby, and it's impossible to list it all because new reports and publications are continually becoming available.

IMPORTANT ADDRESSES

National Association of Rocketry
P.O. Box 177
Altoona, WI 54720
(800) 262-4872
www.nar.org

Aerocon Systems
www.aeroconsystems.com

Aerospace Specialty Products
P.O. Box 1408
Gibsonton, FL 33534
(813) 741-0032
www.asp-rocketry.com

Aerotech, Inc.
2113 W. 850 N. Street
Cedar City, UT 84720
(435) 865-7100
www.aerotech-rocketry.com

Apogee Components
1130 Elkton Drive, Suite A
Colorado Springs, CO 80907
(719) 535-9335
www.apogeerockets.com

B2 Rocketry Company
105 Junco Way
Savannah, GA 31419
(912) 925-1638
www.b2rocketry.com

Balsa Machining Service
11995 Hillcrest Drive
Lemont, IL 60439
(630) 257-5420
www.balsamachining.com

Mike Dorffler Replicas
2418 Greenway Circle
Canon City, CO 81212
mkdorffler@earthlink.net

Estes Industries
1295 H Street
Penrose, CO 81240
(800) 525-7563
www.estesrockets.com

Extreme Rocketry Magazine
3020 Bryant Avenue
Las Vegas, NV 89102
(702) 233-8222
www.extremerocketry.com

FlisKits, Inc.
6 Jennifer Drive
Merrimack, NH 03054
(603) 424-3388
www.fliskits.com

G-Wiz Flight Computers
(see Pratt Hobbies)

Holatron Systems
748 21st Avenue
Honolulu, HI 96816
(808) 372-0956
www.holatron.com

LOC/Precision
P.O. Box 470396
Broadview Heights, OH 44147
(440) 546-0413
www.locprecision.com

Magnum, Inc.
P.O. Box 124
Mechanicsburg, OH 43044
www.magnumrockets.com

Missile Works
453 E. Wonderview Avenue/PMB 184
Estes Park, CO 80517
(303) 823-9222
www.missileworks.com

Perfectflite
15 Pray Street
Amherst, MA 01002
(413) 549-3444
www.perfectflite.com

Pratt Hobbies
2513 Iron Forge Road
Herndon, VA 20171
(703) 689-3541
www.pratthobbies.com

Public Missiles Ltd.
25140 Terra Industrial Drive
Chesterfield Township, MI 48051
1-888-PUBLIC-M
www.publicmissiles.com

Qualified Competition Rockets
7021 Forest View Drive
Springfield, VA 22150
(703) 451-2808
www.cybertravelog.com/qcr

Quest Aerospace, Inc.
6012 East Hidden Valley Drive
Cave Creek, AZ 85331
(800) 858-7302
www.questaerospace.com

Range Pro Static Test Stand
Art Rose
2708 East Hillery
Phoenix, AZ 85032
(602) 404-3686

Saturn Press
P.O. Box 3709
Ann Arbor, MI 48106-3709
(734) 677-2321
petealway@aol.com

SpaceCad
Tubinger Strasse 6, D-71636
Ludwigsburg, Germany
www.spacecad.com

SuperCal Decal System
Micro Format, Inc.
Wheeling, Il 60090

Tango Papa Decal
1901 Mitman Road
Easton, PA 18040
(484) 767-8731
www.tangopapadecals.com

Totally Tubular
10555 McCabe Road
Brighton, MI 48116-8526
(810) 231-3471
www.totubular.com

Transolve
www.transolve.com

Tripoli Rocketry Association
www.tripoli.org

Appendix II

MODEL ROCKET CP CALCULATION

From James S. and Judith A. Barrowman

Nose:

L_N = length of nose

For Cone	For Ogive
$(C_N)_N = 2$	$(C_N)_N = 2$
$X_N = 0.666L_N$	$X_N = 0.466L_N$

Conical Transition (for both increasing and decreasing diameters):

d_F = diameter of front of transition
d_R = diameter of rear of transition
L_T = length of transition piece (distance from d_F to d_R)
X_P = distance from tip of nose to front of transition
d = diameter of base of nose

$$(C_N)_T = 2\left[\left(\frac{d_R}{d}\right)^2 - \left(\frac{d_F}{d}\right)^2\right]$$

NOTE: $(C_N)_T$ will be negative for conical boat tail.

$$X_T = X_P + \frac{L_T}{3}\left[1 + \frac{1 - \frac{d_F}{d_R}}{1 - \left(\frac{d_F}{d_R}\right)^2}\right]$$

Fins (for multistage models, calculate each set of fins separately, using a different X_B):

C_R = fin root chord
C_T = fin tip chord

S = fin semispan
L_F = length of fin midchord line
R = radius of body rear end
X_R = distance between fin root leading edge and fin tip leading edge parallel to body
X_B = distance from nose tip to fin root chord leading edge

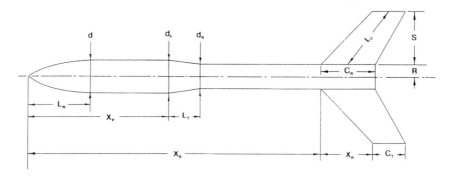

For 3 fins:

$$(C_N)_F = \left[1 + \frac{R}{S+R}\right]\left[\frac{12\left(\frac{S}{d}\right)^2}{1 + \sqrt{1 + \left(\frac{2L_F}{C_R + C_T}\right)^2}}\right]$$

For 4 fins:

$$(C_N)_F = \left[1 + \frac{R}{S+R}\right]\left[\frac{16\left(\frac{S}{d}\right)^2}{1 + \sqrt{1 + \left(\frac{2L_F}{C_R + C_T}\right)^2}}\right]$$

$$\overline{X}_F = X_B + \frac{X_R}{3}\frac{(C_R + 2C_T)}{(C_R + C_T)} + \frac{1}{6}\left[(C_R + C_T) - \frac{(C_R C_T)}{(C_R + C_T)}\right]$$

Total Values:

$$(C_N)_R = (C_N)_N + (C_N)_T + (C_N)_F + \ldots$$

(the sum of the force coefficient C_N of each part calculated)

CP Distance from Nose Tip = $\overline{X}$

$$= \frac{(C_N)_N\overline{X}_N + (C_N)_T\overline{X}_T + (C_N)_F\overline{X}_F}{(C_N)_R}$$

(the sum of the products of the force coefficient C_N and the part CP of each part divided by the total rocket C_N)

ROCKET ALTITUDE SIMULATION: COMPUTER PROGRAM RASP-93

NOTE: The following program is written in BASIC. The particular version of BASIC for your computer may differ slightly in command structure; check your computer's instruction manual.

```
10   REM Program Code "RASP-93"
20   REM Rocket Altitude Simulation Program
30   REM by G. Harry Stine
40   REM Revision RASP-90M by Timothy Barr, Falingtrea Enterprises, Sept 1991
50   REM Revision RASP-93 by G. Harry Stine, May 1993
60   FMT.1$ = "###.# ####.#### ###.#### ###.#### ##.####"
70   TIME = 0
80   VEL = 0
90   MAX.VEL = 0
100  ALTI = 0
110  TIME.INC = .1
120  PI = 3.14159
130  RHO = 1.2062
140  G.FORCE = 9.81001
150  PRINT "Is there a printer available? (Y or N)";
160  INPUT TEST$
170  IF((LEFT$(TEST$,1) = "N") OR (LEFT$(TEST$,1) = "n")) THEN PRINT.FLAG = 0:GOTO 200
180  PRINT.FLAG = -1
190  OPEN "O", #1, "LPT1:"
200  PRINT "Enter rocket no-engine (empty) weight:"
210  INPUT EMPTY.WEIGHT
220  GOSUB 1230
230  PRINT "Enter the Rocket's maximum body tube diameter:"
240  INPUT BT.RADIUS
250  GOSUB 1140
260  BT.RADIUS = BT.RADIUS/2
270  PRINT "Enter the Drag Coefficient:"
280  INPUT C.D
290  PRINT "Enter a Rocket Motor type code (i.e. B4):"
300  INPUT RM.CODE$
310  IF (RM.CODE$ = "A6") OR (RM.CODE$ = "a6") THEN RESTORE 1370: GOTO 620
320  IF (RM.CODE$ = "A8") OR (RM.CODE$ = "a8") THEN RESTORE 1390: GOTO 620
330  IF (RM.CODE$ = "B4") OR (RM.CODE$ = "b4") THEN RESTORE 1410: GOTO 620
340  IF (RM.CODE$ = "B6") OR (RM.CODE$ = "b6") THEN RESTORE 1440: GOTO 620
350  IF (RM.CODE$ = "B8") OR (RM.CODE$ = "b8") THEN RESTORE 1460: GOTO 620
360  IF (RM.CODE$ = "C5") OR (RM.CODE$ = "c5") THEN RESTORE 1480: GOTO 620
```

```
370  IF (RM.CODE$ = "C6") OR (RM.CODE$ = "c6") THEN RESTORE 1510: GOTO 620
380  IF (RM.CODE$ = "D12") OR (RM.CODE$ = "d12") THEN RESTORE 1540: GOTO 620
390  IF (RM.CODE$ = "D13") OR (RM.CODE$ = "d13") THEN RESTORE 1560: GOTO 620
400  IF (RM.CODE$ = "D21") OR (RM.CODE$ = "d21") THEN RESTORE 1610: GOTO 620
410  IF (RM.CODE$ = "E15") OR (RM.CODE$ = "e15") THEN RESTORE 1640: GOTO 620
420  IF (RM.CODE$ = "E18") OR (RM.CODE$ = "e18") THEN RESTORE 1680: GOTO 620
430  IF (RM.CODE$ = "E25") OR (RM.CODE$ = "e25") THEN RESTORE 1730: GOTO 620
440  IF (RM.CODE$ = "E30") OR (RM.CODE$ = "e30") THEN RESTORE 1760: GOTO 620
450  IF (RM.CODE$ = "E45") OR (RM.CODE$ = "e45") THEN RESTORE 1790: GOTO 620
460  IF (RM.CODE$ = "E50") OR (RM.CODE$ = "e50") THEN RESTORE 1820: GOTO 620
470  IF (RM.CODE$ = "F24") OR (RM.CODE$ = "f24") THEN RESTORE 1840: GOTO 620
480  IF (RM.CODE$ = "F25") OR (RM.CODE$ = "f25") THEN RESTORE 1890: GOTO 620
490  IF (RM.CODE$ = "F32") OR (RM.CODE$ = "f32") THEN RESTORE 1940: GOTO 620
500  IF (RM.CODE$ = "F44") OR (RM.CODE$ = "f44") THEN RESTORE 1980: GOTO 620
510  IF (RM.CODE$ = "F50") OR (RM.CODE$ = "f50") THEN RESTORE 2020: GOTO 620
520  IF (RM.CODE$ = "F55") OR (RM.CODE$ = "f55") THEN RESTORE 2060: GOTO 620
530  IF (RM.CODE$ = "F72") OR (RM.CODE$ = "f72") THEN RESTORE 2090: GOTO 620
540  IF (RM.CODE$ = "F80") OR (RM.CODE$ = "f80") THEN RESTORE 2120: GOTO 620
550  IF (RM.CODE$ = "G25") OR (RM.CODE$ = "g25") THEN RESTORE 2150: GOTO 620
560  IF (RM.CODE$ = "G40") OR (RM.CODE$ is "g40") THEN RESTORE 2220: GOTO 620
570  IF (RM.CODE$ = "G42") OR (RM.CODE$ = "g42") THEN RESTORE 2270: GOTO 620
580  IF (RM.CODE$ = "G55") OR (RM.CODE$ = "g55") THEN RESTORE 2310: GOTO 620
590  IF (RM.CODE$ = "G80") OR (RM.CODE$ = "g80") THEN RESTORE 2350: GOTO 620
600  PRINT "Did not recognize the rocket motor code requested."
610  PRINT "Please try again.": GOTO 290
620  READ MOTOR.MASS, MASS.DEC, DELAY, M.FORCE
630  ROCKET.MASS = ROCKET.MASS + MOTOR.MASS
640  DRAG.VAL = .5 * RHO * PI * C.D * BT.RADIUS ^ 2
650  IF NOT PRINT.FLAG THEN GOTO 720
660  PRINT#1, "The total weight of the rocket is "; ROCKET.MASS*1000; " grams."
670  PRINT #1, "The weight of the motor is "; MOTOR.MASS * 1000; " grms."
680  PRINT #1, "The Maximum Body Tube Radius is "; BT.RADIUS * 1000; " mm."
690  PRINT #1, "The Coefficient of drag used is "; C.D
700  PRINT #1,
710  PRINT #1,
720  LINE.COUNT = 7
730  PRINT " Time  Altitude  Velocity  Acceleration  Weight"
740  IF (NOT PRINT.FLAG) THEN GOTO 760
750  PRINT #1, " Time  Altitude  Velocity  Acceleration  Weight"
760  REM Altitude calculation loop starts here.
770  IF (NOT PRINT.FLAG) OR (LINE.COUNT MOD 60 < > 0) THEN GOTO 800
780  PRINT #1, CHR$(12) ' Replace with code(s) to advance printer page
790  PRINT #1, " Time  Altitude  Velocity  Acceleration  Weight"
800  IF (LINE.COUNT MOD 20) < > 0 THEN GOTO 850
810  PRINT "Press ENTER to Continue";
820  INPUT JUNK
830  PRINT " Time  Altitude  Velocity  Acceleration  Weight"
840  LINE.COUNT = LINE.COUNT + 1
850  TIME = TIME + TIME.INC
860  ACCEL = M.FORCE/ROCKET.MASS - G.FORCE - DRAG.VAL *VEL^2/ROCKET.MASS
870  VEL = VEL + ACCEL * TIME.INC
880  IF VEL < 0 THEN GOTO 980
890  ALTI = ALTI + VEL * TIME.INC
900  IF M.FORCE > 0 THEN ROCKET.MASS = ROCKET.MASS - MASS.DEC
910  IF M.FORCE > 0 THEN READ M.FORCE
920  IF VEL > MAX.VEL THEN MAX.VEL = VEL
930  PRINT USING FMT.1&; TIME; ALTI; VEL; ACCEL; ROCKET.MASS
940  LINE.COUNT = LINE.COUNT + 1
950  IF NOT PRINT.FLAG THEN GOTO 760
960  PRINT #1, USING FMT.1&; TIME; ALTI; VEL; ACCEL; ROCKET.MASS
970  GOTO 760
980  PRINT "Maximum Altitude attained ="; ALTI; "meters."
990  PRINT "or"; ALTI * 3.28084; "feet."
1000  PRINT "Time to peak altitude ="; TIME; "seconds."
1010  DELAY = TIME - DELAY
```

```
1020   PRINT "Recommended delay time for model ="; DELAY; "seconds."
1030   PRINT "Maximum velocity attained ="; MAX.VEL; " meters per second."
1040   IF NOT PRINT.FLAG THEN STOP
1050   PRINT #1, "The Maximum Altitude attained was "; ALTI; "meters,";
1060   PRINT#1, " or"; ALTI *3.2808; "feet."
1070   PRINT #1, "The time to peak altitude was "; TIME; " seconds."
1080   PRINT#1, "Recommended time delay for the ";RM.CODE&;" motor used is:;
1090   PRINT #1, DELAY; " seconds."
1100   PRINT #1, "Maximum velocity attained was "; MAX.VEL; " meters/second,";
1110   PRINT #1, "or"; MAX.VEL * 3.2808; " feet/second."
1120   PRINT#1, CHR&(12)
1130   STOP
1140   REM Subroutine that translates length units
1150   PRINT "Choose the length measurement unit used:"
1160   PRINT " 1 - inches  2 - millimeters  3 - feet  4 - meters"
1170   INPUT INDEX
1180   IF (INDEX < 1) OR (INDEX > 4) THEN PRINT "INVALID! Try again": GOTO 1150
1190   IF INDEX < 2 THEN BT.RADIUS = BT.RADIUS * .0254: GOTO 1220
1200   IF INDEX < 3 THEN BT.RADIUS = BT.RADIUS * .001: GOTO 1220
1210   IF INDEX < 4 THEN BT.RADIUS = BT.RADIUS * .3048: GOTO 1220
1220   RETURN
1230   REM Subroutine that translates mass units
1240   PRINT "Choose the mass measurement unit used:"
1250   PRINT " 1 - ounces  2 - grams  3 - pounds  4 - kilograms"
1260   INPUT INDEX
1270   IF (INDEX < 1) OR (INDEX > 4) THEN PRINT "INVALID! Try again": GOTO 1240
1280   IF INDEX <2 THEN ROCKET.MASS = EMPTY.WEIGHT*2.834952E-02: GOTO 1320
1290   IF INDEX <3 THEN ROCKET.MASS = EMPTY.WEIGHT*.001: GOTO 1320
1300   IF INDEX <4 THEN ROCKET.MASS = EMPTY.WEIGHT*.45359237#: GOTO 1320
1310   ROCKET.MASS = EMPTY.WEIGHT
1320   RETURN
1330   REM Data for rocket motors will start here.
1340   REM Format is: Motor mass, mass dec, total burn time,
1350   REM     M.force(@ 0.1 sec)… to M.force(@ burnout)
1360   REM Quest A6 series
1370   DATA 0.0153, 0.0008536, 0.40, 5.91, 11.82, 5.91, 0
1380   REM Estes A8 series
1390   DATA 0.0165, 0.000624, 0.5, 5.2, 10.4, 4, 3.2, 0
1400   REM Estes B4 series
1410   DATA 0.0202, 0.000694, 1.2, 5.2, 10.4, 4, 3.2, 3.2, 3.2, 3.2, 3.2
1420   DATA 3.2, 3.2, 3.2, 0
1430   REM Estes and Quest B6 series
1440   DATA 0.02, 0.00078, 0.8, 6.5, 13, 6, 5, 5, 5, 5, 0
1450   REM Estes B8 series
1460   DATA 0.0193, 0.00104, .6, 11.5, 23, 4, 3, 2, 0
1470   REM Estes C5 series
1480   DATA 0.0255, 0.000747, 1.7, 11.5, 23, 4, 4, 4, 4, 4, 4, 4, 4, 4, 4
1490   DATA 4, 4, 4, 4, 0
1500   REM Estes and Quest C6 series
1510   DATA 0.026, 0.000734, 1.7, 6.5, 13, 6, 5, 5, 5, 5, 5, 5, 5, 5, 5
1520   DATA 5, 5, 5, 5, 0
1530   REM Estes D12 series
1540   DATA 0.043, 0.001466, 1.7, 13, 26, 12, 10, 10, 10, 10, 10, 10, 10
1550   DATA 10, 10, 10, 10, 10, 10, 0
1560   REM Aerotech 18mm reloadable D13 series
1570   DATA 0.033, 0.000544, 1.8, 23.13, 22.95, 22.02, 20.91, 19.79, 18.01
1580   DATA 16.32, 14.68, 12.23, 9.79, 6.89, 4.89, 3.11, 2.00, 1.11, 0.80
1590   DATA 0.44, 0
1600   REM Aerotech D21 series
1610   DATA 0.0236, 0.00096, 1.0, 30.08, 30.26, 30.08, 28.03, 25.36, 21.36
1620   DATA 16.91, 12.46, 5.43, 0
1630   REM Aerotech E15 series
1640   DATA 0.048, 0.000703, 3.0, 28.92, 24.03, 22.69, 20.91, 20.02, 19.58
1650   DATA 19.13, 18.69, 18.24, 17.8, 17.35, 16.91, 16.46, 16.02, 15.13
1660   DATA 14.68, 13.79, 12.46, 11.57, 10.68, 9.79, 8.9, 7.56, 5.78
```

```
1670   DATA 4.45, 3.11, 2.22, 1.74, 1.33, 0
1680   REM Aerotech 24mm Reloadable E18 series
1690   DATA .057, 0.0009409, 2.2, 31.58, 31.58, 31.36, 31.14, 30.69, 30.25
1700   DATA 29.36, 28.02, 26.24, 24.02, 21.80, 19.57, 16.91, 13.34, 11.12
1710   DATA 8.45, 5.34, 3.56, 2.22, 1.33, 0.44, 0
1720   REM Aerotech E25 series
1730   DATA 0.0249, 0.00122, 0.9, 32.48, 32.93, 32.22, 31.59, 29.81, 25.81
1740   DATA 20.91, 14.24, 0
1750   REM Aerotech E30 series
1760   DATA 0.0468, 0.0014846, 1.3, 49.84, 50.11, 48.95, 46.72, 43.61, 40.94
1770   DATA 37.82, 32.04, 25.81, 15.57, 6.23, 2.31, 0
1780   REM Aerotech E45 series
1790   DATA 0.0369, 0.0024, 0.8, 71.2, 79.65, 82.68, 78.32, 71.20, 13.35
1800   DATA 3.56, 0
1810   REM Aerotech E50 series
1820   DATA 0.0384, 0.0023, 0.8, 41.38, 52.06, 58.21, 61.41, 63.19, 63.19
1830   DATA 60.52, 0
1840   REM Aerotech 24mm Reloadable F24 series
1850   DATA 0.062, 0.00115, 2.2, 39.14, 39.14, 39.14, 38.70, 38.25, 37.59
1860   DATA 36.70, 35.36, 33.80, 31.14, 28.02, 24.46, 21.35, 17.35, 12.45
1870   DATA 8.90, 5.78, 3.56, 2.22, 1.33, 0.44, 0
1880   REM Aerotech F25 series
1890   DATA .089, 0.001297, 3.2, 50.91, 45.57, 46.28, 47.17, 46.28, 44.05
1900   DATA 41.83, 40.05, 37.82, 35.6, 34.26, 32.04, 30.7, 28.92, 27.14, 26.25
1910   DATA 24.92, 23.58, 21.8, 20.47, 18.69, 16.91, 15.13, 13.35, 11.12, 8.9
1920   DATA 4.45, 2.67, 1.78, 0.89, 0.44, 0
1930   REM Aerotech F32 series
1940   DATA 0.072, 0.001508, 2.5, 50.28, 43.61, 42.72, 41.83, 41.83, 40.94
1950   DATA 40.94, 40.05, 39.6, 39.16, 38.71, 38.27, 37.82, 37.38, 36.93,
1960   DATA 35.6, 34.71, 32.48, 30.26, 24.46, 16.46, 8.45, 5.78, 2.67, 0
1970   REM Aerotech F44 series
1980   DATA 0.0735, 0.002022, 1.8, 64.52, 62.3, 60.52, 59.18, 57.76, 56.07
1990   DATA 55.18, 53.84, 52.95, 51.17, 49.84, 48.5, 46.72, 40.05, 24.47
2000   DATA 11.12, 5.78, 0
2010   REM Aerotech F50 series
2020   DATA 0.083, 0.0022529, 1.7, 82.68, 79.21, 78.76, 75.2, 71.64, 67.64
2030   DATA 63.19, 58.74, 54.29, 48.5, 42.27, 34.26, 24.03, 11.57, 4.45
2040   DATA 3.56, 0
2050   REM Aerotech F55 series
2060   DATA 0.0457, 0.002667, 0.9, 93.45, 96.79, 95.67, 91.67, 73.42
2070   DATA 28.92, 14.24, 5.78, 0
2080   REM Aerotech F72 series
2090   DATA 0.0731, 0.003346, 1.1, 103.68, 107.24, 109.47, 109.47, 108.58
2100   DATA 105.91, 84.1, 43.16, 21.36, 7.12, 0
2110   REM Aerotech F80 series
2120   DATA 0.085, 0.00381, 1, 63.19, 74.31, 84.1, 92.11, 99.23, 105.91
2130   REM Aerotech G40 series
2140   REM Aerotech G25 series
2150   DATA 0.105, 0.0011417, 4.8, 31.59, 37.38, 40.49, 34.71, 33.37, 33.82
2160   DATA 34.71, 35.15, 36.04, 36.04, 36.49, 36.93, 36.93, 36.93, 37.38
2170   DATA 37.38, 37.38, 36.93, 36.93, 36.93, 36.04, 35.15, 34.26, 33.37
2180   DATA 32.04, 30.7, 28.92, 27.14, 24.92, 22.96, 21.8, 20.02, 18.69
2190   DATA 16.91, 15.13, 13.79, 12.46, 10.23, 9.34, 8.45, 7.12, 5.78, 4
2200   DATA 2.67, 2.22, 1.33, 0.89, 0
2210   REM Aerotech G40 series
2220   DATA 0.125, 0.00208, 3, 67.64, 55.62, 52.06, 50.73, 49.84, 48.95
2230   DATA 48.06, 47.17, 46.72, 46.28, 45.39, 44.5, 43.61, 42.72, 41.83
2240   DATA 41.28, 40.94, 40.05, 39.16, 38.27, 37.38, 35.6, 34.71, 33.82
2250   DATA 32.04, 30.26, 26.7, 22.52, 16.02, 0
2260   REM Aerotech G42 series
2270   DATA 0.0908, 0.0023238, 2.1, 66.75, 66.75, 66.75, 66.75, 66.75, 66.3
2280   DATA 65.41, 63.19, 60.52, 57.85, 53.84, 50.73, 45.83, 37.38, 27.59
2290   DATA 17.8, 8.9, 5.34, 4, 1.78, 0
2300   REM Aerotech G55 series
2310   DATA 0.1037, 0.0027174, 2.3, 101.28, 88.11, 84.55, 84.1, 82.32, 79.65
```

```
2320   DATA 76.54, 73.42, 69.42, 66.75, 64.52, 62.74, 60.52, 56.51, 45.39
2330   DATA 31.15, 22.69, 17.35, 13.79, 11.12, 7.56, 4.45, 0
2340   REM Aerotech G80 series
2350   DATA 0.12, 0.0038267, 1.5, 112.58, 113.47, 112.58, 111.25, 109.47
2360   DATA 106.18, 102.35, 98.79, 93.89, 85.44, 72.09, 48.95, 22.25, 10.68, 0
2370   END
```

STATIC STABILITY CALCULATION: COMPUTER PROGRAM STABCAL-2

NOTE: The following program is written in BASIC. The particular version of BASIC for your computer may differ slightly in command structure; check your computer's instruction manual.

```
10 REM "STABCAL-2" PROGRAM
20 REM ROCKET SUBSONIC STATIC STABILITY CALCULATIONS
30 REM by G. Harry Stine
40 REM Copyright (C) 1991 by G. Harry Stine
50 REM May be copied and used freely but not sold.
60 REM Revised from "STABCALC-1" written in 1979.
70 REM
80 PRINT "ROCKET STATIC STABILITY CALCULATIONS"
90 PRINT "Rocket name:"
91 INPUT N$
100 PRINT "Dimensional system used, inches or mm?"
101 INPUT P$
110 PRINT "ROCKET DIMENSIONS:"
120 PRINT "Nose length:"
130 INPUT X1
140 PRINT "Nose base diameter:"
150 INPUT D1
160 PRINT "Any transitions?"
170 INPUT A$
180 IF A$ = "NO" GOTO 490
190 PRINT "Transition 1 front diameter:"
200 INPUT D2
210 PRINT "Transition 1 rear diameter:"
220 INPUT D3
230 PRINT "Transition 1 length:"
240 INPUT L1
250 PRINT "Distance, nose tip to Trans.1 front:"
260 INPUT X2
270 PRINT "Any more transitions?"
280 INPUT B$
290 IF B$ = "NO" GOTO 490
300 PRINT "Transition 2 front dia.:"
310 INPUT D4
320 PRINT "Transition 2 rear dia.:"
330 INPUT D5
340 PRINT "Transition 2 length:"
```

331

```
350 INPUT L2
360 PRINT "Distance, nose tip to Trans.2 front:"
370 INPUT X3
380 PRINT "Any more transitions?"
390 INPUT C$
400 IF C$ = "NO" GOTO 490
410 PRINT "Transition 3 front dia.:"
420 INPUT D6
430 PRINT "Transition 3 rear dia.:"
440 INPUT D7
450 PRINT "Transition 3 length:"
460 INPUT L3
470 PRINT "Distance, nose tip to Trans.3 front:"
480 INPUT X4
490 PRINT "Fin 1 root chord:"
500 INPUT C1
510 PRINT "Fin 1 tip chord:"
520 INPUT C2
530 PRINT "Fin 1 semi-span:"
540 INPUT S1
550 PRINT "Fin 1 mid-chord line length:"
560 INPUT S2
570 PRINT "Number of fins, Fin 1:"
580 INPUT A1
590 PRINT "Radius of body at Fin 1:"
600 INPUT R1
610 PRINT "Fin 1 root to tip LE sweep distance:"
620 INPUT X5
630 PRINT "Distance, nose tip to Fin 1 root LE:"
640 INPUT X6
650 PRINT "Any more fins?"
660 INPUT D$
670 IF D$ = "NO" GOTO 1030
680 PRINT "Fin 2 root chord:"
690 INPUT C3
700 PRINT "Fin 2 tip chord:"
710 INPUT C4
720 PRINT "Fin 2 semi-span:"
730 INPUT S3
740 PRINT "Fin 2 mid-chord length:"
750 INPUT S4
760 PRINT "Number of fins, Fin 2:"
770 INPUT A2
780 PRINT "Radius of body at Fin 2:"
790 INPUT R2
800 PRINT "Fin 2 root to tip LE sweep distance:"
810 INPUT X7
820 PRINT "Distance, nose tip to Fin 2 root LE:"
830 INPUT X8
840 PRINT "Any more fins?"
850 INPUT E$
860 IF E$ = "NO" GOTO 1030
870 PRINT "Fin 3 root chord:"
880 INPUT C5
890 PRINT "Fin 3 tip chord:"
900 INPUT C6
910 PRINT "Fin 3 semi-span:"
920 INPUT S5
930 PRINT "Fin 3 mid-chord length:"
940 INPUT S6
950 PRINT "Number of fins, Fin 3:"
960 INPUT A3
970 PRINT "Radius of body at Fin 3:"
980 INPUT R3
990 PRINT "Fin 3 root to tip LE sweep distance:"
```

```
1000 INPUT X9
1010 PRINT "Distance, nose tip to Fin 3 root LE:"
1020 INPUT X10
1030 REM
1040 REM
1050 PRINT "CALCULATION OF NOSE:"
1060 PRINT "If nose shape is ogive, enter 1; cone 2; parabola 3"
1070 INPUT U
1080 IF U = 1 GOTO 1110
1090 IF U = 2 GOTO 1130
1100 IF U = 3 GOTO 1150
1110 N2 = X1 *.466
1120 GOTO 1160
1130 N2 = X1 *.666
1140 GOTO 1160
1150 N2 = X1 *.5
1160 N1 = 2
1170 PRINT "Nose normal force = "; N1
1180 PRINT "Nose CP = "; N2
1190 IF A$ = "NO" GOTO 1600
1200 REM CALCULATION OF TRANSITION 1 ****************
1210 PRINT "CALCULATION OF TRANSITION NO. 1:"
1220 T1 = (D2/D1) *(D2/D1)
1230 T2 = (D3/D1) *(D3/D1)
1240 T3 = (T2-T1)/2
1250 T4 = (D2/D3) *(D2/D3)
1260 T5 = 1-(D2/D3)
1270 T6 = 1-(T4)
1280 T7 = 1 + (T5/T6)
1290 T8 = (L1/3) *T7
1300 T9 = X2 + T8
1310 PRINT "Transition 1 normal force: "; T3
1315 PRINT "Transition 1 CP: "; T9
1320 IF B$ = "NO" GOTO 1600
1330 REM Calculation of Transition 2 ****************
1340 PRINT "CALCULATION OF TRANSITION NO. 2:"
1350 T11 = (D4/D1) *(D4/D1)
1360 T12 = (D5/D1) *(D5/D1)
1370 T13 = (T12-T11) *2
1380 T14 = (D4/D5) *(D4/D5)
1390 T15 = 1-(D4/D5)
1400 T16 = 1-T14
1410 T17 = 1 + (T15/T16)
1420 T18 = (L2/3) *T17
1430 T19 = X3 + T18
1440 PRINT "Transition 2 normal force: "; T13
1450 PRINT "Transition 2 CP: "; T19
1460 IF C$ = "NO" GOTO 1600
1470 REM Calculation of Transition 3 ***************
1480 PRINT "CALCULATION OF TRANSITION NO. 3:"
1490 T21 = (D6/D1) *(D6/D1)
1500 T22 = (D7/D1) *(D7/D1)
1510 T23 = (T22-T21) *2
1520 T24 = (D6/D7) *(D6/D7)
1530 T25 = 1-(D6/D7)
1540 T26 = 1-T14
1550 T27 = 1 + (T25/T26)
1560 T28 = (L3/3) *T27
1570 T29 = X4 + T28
1580 PRINT "Transition 3 normal force: "; T23
1590 PRINT "Transition 3 CO: "; T29
1600 REM Calculation of Fin 1 ***********************
1610 PRINT "CALCULATION OF FIN NO. 1:"
1620 F1 = (S1/D1) *(S1/D1)
1630 IF A1 = 3 THEN 1650
```

```
1640 IF A1 = 4 THEN 1670
1650 F2 = F1 *13.85
1660 GOTO 1680
1670 F2 = F1 *16
1680 F3 = 2 *S2
1690 F4 = C1 + C2
1700 F5 = (F3/F4) *(F3/F4)
1710 F6 = 1 + F5
1720 F7 = SQR(F6)
1730 F8 = 1 + F7
1740 F9 = F2/F8
1750 F10 = S1 + R1
1760 F11 = R1/F10
1770 F12 = 1 + F11
1780 F13 = F12 *F9
1790 F14 = X5/3
1800 F15 = S *C2
1810 F16 = F15 + C1
1820 F17 = F16/F4
1830 F18 = F14 *F17
1840 F19 = C1 *C2
1850 F20 = F19/F4
1860 F21 = F4-F20
1870 F22 = F21/6
1880 F23 = X6 + F18 + F22
1890 PRINT "Fin 1 normal foce: "; F13
1900 PRINT "Fin 1 CP: "; F23
1910 IF D$ = "NO" GOTO 2550
1920 REM Calculation of Fin 2 **********************
1930 PRINT "CALCULATION OF FIN NO. 2:"
1940 F31 = (S4/D1) *(S4/D1)
1950 IF A2 = 3 GOTO 1970
1960 IF A2 = 4 GOTO 1990
1970 F32 = F31 *13.85
1980 GOTO 2000
1990 F32 = F31 *16
2000 F33 = 2 *S4
2010 F34 = C3 + C4
2020 F35 = (F33/F34) *(F33/F34)
2030 F36 = 1 + F35
2040 F37 = SQR(F36)
2050 F38 = 1 + F37
2060 F39 = F32/F38
2070 F40 = S3 + R2
2080 F41 = R2/F40
2090 F42 = 1 + F41
2100 F43 = F42 *F39
2110 F44 = X7/3
2120 F45 = 2 *C4
2130 F46 = F45 + C3
2140 F47 = F46/F34
2150 F48 = F44 *F47
2160 F49 = C3 *C4
2170 F50 = F49/F34
2180 F51 = F34-F50
2190 F52 = F51/6
2200 F53 = X8 + F48 + F52
2210 PRINT "Fin 2 normal force: "; F43
2220 PRINT "Fin 2 CP: "; F53
2230 IF E$ = "NO" GOTO 2550
2240 REM Calculation of Fin 3 **********************
2250 PRINT "CALCULATION OF FIN NO. 3:"
2260 F61 = (S5/D1) *S5/D1)
2270 IF A3 = 3 GOTO 2290
2280 IF A3 = 4 GOTO 2310
```

```
2290 F62 = F61 *13.85
2300 GOTO 2320
2310 F62 = F61 *16
2320 F63 = 2 *S6
2330 F64 = C5 + C6
2340 F65 = (F63/F64) *(F63/F64)
2350 F66 = 1 + F65
2360 F67 = SQR(F66)
2370 F68 = 1 + F67
2380 F69 = F62/F68
2390 F70 = S5 + R3
2400 F71 = R3 + F70
2410 F72 = 1 + F71
2420 F73 = F72 *F69
2430 F74 = X9/3
2440 F75 = 2 *S6
2450 F76 = F75 + C5
2460 F77 = F76/F64
2470 F78 = F74 *F77
2480 F79 = C5 *C6
2490 F80 = F79/F64
2500 F81 = F64-F80
2510 REM
2520 F83 = X10 + F78 + F82
2530 PRINT "Fin 3 normal force: "; F73
2540 PRINT "Fin 3 CP: "; F83
2550 REM Calculation of total rocket ****************
2560 PRINT "CALCULATION OF TOTAL ROCKET:"
2570 REM
2580 M1 = N1 + T3 + T13 + T23 + F13 + F43 + F73
2590 PRINT "Total rocket normal force = "; M1
2600 M2 = N1 *N2
2610 PRINT "Nose moment = "; M2
2620 M3 = T3 *T9
2630 PRINT "Transition 1 moment = "; M3
2640 M4 = T13 *T19
2650 PRINT "Transition 2 moment = "; M4
2660 M5 = T23 *T29
2670 PRINT "Transition 3 moment = "; M5
2680 M6 = F13 *F23
2690 PRINT "Fin 1 moment = "; M6
2700 M7 = F43 *F53
2710 PRINT "Fin 2 moment = "; M7
2720 M8 = F73 *F83
2730 PRINT "Fin 3 moment = "; M8
2740 M9 = M2 + M3 + M4 + M5 + M6 + M7 + M8
2750 PRINT "Total rocket moment = "; M9
2760 X12 = M9/M1
2770 PRINT "CP is"; X12; "dimensional units behind nose tip."
2780 END
```

THE TRIPLE-TRACK TRACKER

by Arthur H. "Trip" Barber, NAR 4322
© 1979 by the National Association of Rocketry
Reprinted with permission

Building a pair of Triple-Track Trackers requires approximately six hours of work and less than $20 worth of materials. Triple-Track Trackers may be mounted on any sturdy tripods having 1/4" bolt fittings on the top. Standard photo tripods are ideal. Zeroing in azimuth is done by rotating the tracker plate on the tripod. Elevation zeroing is accomplished by adjusting the springloaded bolts on each corner of the tracker plate and by adjusting the tripod legs.

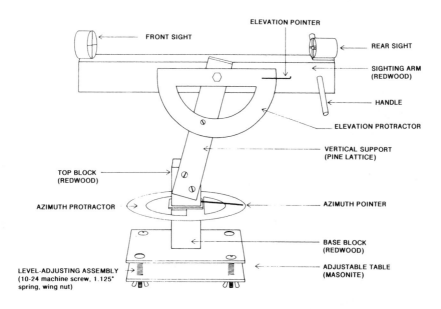

PARTS LIST

(Quantities in Parentheses)

(2) 6″ 180° plastic protractors

(2) 6″ 360° plastic protractors

(1) 36″ length 1 1/2″ × 1 1/2″ clear redwood

(1) 12″ × 12″ piece 1/8″ tempered Masonite

(8) 1.125″ × 0.375″ compression springs

(2) 0.750″ × 0.375″ compression springs

(1) 16″ length 1/4″ hardwood dowel

(1) 14″ length 1/4″ × 1 1/8″ pine lattice

(1) 8″ length 1/2″ brass tubing

(1) 8 1/2″ length 15/32″ brass tubing

(1) 1 1/2″ × 3″ piece 1/16″ rubber gasket material

(1) 1 1/2″ length of BT-55 body tube

(1) 2″ length BT-50 body tube

(1) 1″ length 1/8″ launch lug

(6) 3/4″ metal washers

(6) 1/2″ × 4 flat head wood screws

(8) 2″ 10-24 round head machine screws

(8) 10-24 wing nuts

(2) 3″ 1/4″ hex head machine screws

(4) 1/4″ wing nuts

(4) 1″ × 8 round head wood screws

(4) 3/4″ × 4 flat head wood screws

(1) 12″ length 0.47″ music wire

(1) 3/4″ × 2″ piece hard 3/32″ balsa

(1) 3/8″ × 1″ piece hard 3/16″ balsa

ASSEMBLY INSTRUCTIONS

NOTE: The italicized parts of these instructions are particularly critical to the accuracy of the tracker. Do them *exactly* as specified.

1. Cut two 6″ squares of Masonite.
2. Place the two squares on top of each other and drill one 13/64″ diameter hole in each corner equidistant from the edges and about 1″ in from each corner.
3. Locate the center of each square by drawing diagonals between opposite corners.

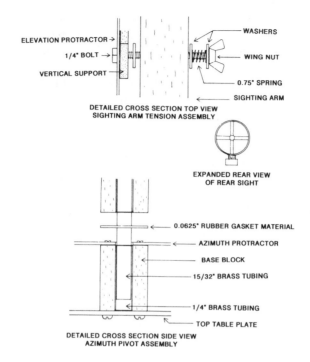

ELEVATION PROTRACTOR →

1/4" BOLT →

VERTICAL SUPPORT →

WASHERS

WING NUT

0.75" SPRING

SIGHTING ARM

DETAILED CROSS SECTION TOP VIEW
SIGHTING ARM TENSION ASSEMBLY

EXPANDED REAR VIEW
OF REAR SIGHT

0.0625" RUBBER GASKET MATERIAL

AZIMUTH PROTRACTOR

BASE BLOCK

15/32" BRASS TUBING

1/4" BRASS TUBING

TOP TABLE PLATE

DETAILED CROSS SECTION SIDE VIEW
AZIMUTH PIVOT ASSEMBLY

4. On the top square, drill two 3/16" holes along one diagonal, each 1/2" on either side of the center.

5. On the bottom square, drill a 1/4" hole in the center.

6. Cut two 2" lengths and one 13" length of the 1 1/2" square redwood. *The cut faces must be flat and perpendicular to the sides.* Use either a power saw or a fine-tooth hand saw and a miter.

7. Locate the center of one 1 1/2" square face on each 2" redwood piece.

8. Place each 2" redwood piece in a tightly clamped holder and *using a drill press,* drill a 1/4" hole through the entire length at the center of the marked square face. Then drill a 1/2" hole with a twist drill bit (not a wood bit) using the 1/4" hole as a guide. *This hole must be perpendicular to the face.*

9. Glue one 2" redwood piece so that the 1/2" hole in it is centered over the center point of the top Masonite square. Set aside.

10. *Using a drill press,* drill 1/4" holes through the width of the 13" length of redwood at points 3/4" and 6" from one end midway between the top and bottom edges of a side. *These holes must be perpendicular to the sides.*

11. Cut a 7" length of pine lattice.

12. *Using a drill press,* drill a 1/4" hole 3/4" from one end of the pine lattice, centered between its edges. Drill two 7/64" holes in the first 1" of the other end. Glue the first 1 1/2" of this same end to the remaining 2" redwood piece so that the long dimension of the pine makes a 15° angle with the axis of the hole in the redwood.

13. Cut two 2″ lengths of the 1/2″ brass tubing and deburr the inside and outside of each end. Roughen the outside of each with sandpaper and epoxy one in the hole of each 2″ redwood piece. Make sure no epoxy gets inside the tubing.

14. When the glue is dry, turn over the top piece of Masonite and drill 9/64″ starter holes for the 1″ × 8 wood screws into the redwood. Screw in the screws.

15. When the glue is dry on the pine that is attached to the second red-wood block, drill 3/32″ starter holes through the holes already in the pine and into the redwood. Screw in the 3/4″ × 4 wood screws.

16. Glue the 8″ length of 1/4″ dowel into the hole nearest one end of the 13″ length of redwood. 3 1/4″ of dowel should stick out on each side.

17. Drill two 1/8″ holes in the 360° protractor in positions where they will be over the lower redwood block after the following step.

18. Epoxy the 360° protractor to the lower 2″ redwood block *so that its center is exactly over the center of the 1/2″ tubing (within 1/64″)*. It may be necessary to carve away a semicircle of material from the pro-tractor (depending on its design) so that the tubing is not blocked. If so, do this before epoxying it down.

19. When the epoxy on the 360° protractor cures, drill 5/64″ starter holes for the 1/2″ × 4 flat head wood screws into the redwood through the holes in the protractor. Screw in the screws.

20. Drill a 1/4″ hole through the center point of the 180° protractor and a 7/64″ hole above the left-hand 75° mark (counting counterclock-wise when the flat edge of the protractor is up).

21. Insert a 3″ × 1/4″ × 20 hex head bolt through the 1/4″ hole in the pro-tractor, then through the 1/4″ hole in the pine lattice. Rotate the pro-tractor until its top (flat) edge is parallel to the top surface of the redwood block. Mark the pine at the point where the small hole in the protractor lies.

22. Drill a 5/64″ starter hole in the pine at the marked point and screw a 1/2″ × 4 wood screw through the hole in the protractor into the pine. Do not let its tip stick out the other side of the pine. Use a little epoxy to further secure the protractor to the pine.

23. Cut two 3/4″ lengths of BT-55 (or equivalent body tube) and two 1″ lengths of BT-50 (or equivalent body tube). Slice one piece of each lengthwise.

24. Cut a 1″ length of 1/8″ launch lug, four 1″ × 3/8″ strips of 3/32″ hard balsa, and one 1″ × 3/8″ strip of 3/16″ hard balsa.

25. Glue the strips of 3/32″ balsa along the launch lug every 90° as if they were fins. Try to keep everything straight so that the assembly looks like a 90° cross when viewed from one end. Set aside to dry.

26. Mark one end of the unsliced piece of BT-55 every 90°. Make a crosshair by taping one end of a piece of dark shroud line or strong thread to the outside of this tube at one mark, then running it across the tube to the opposite mark and taping it down. Do this twice. The

threads should cross at the center of the tube. Trim off the excess thread.

27. Coat the outside of each unsliced body tube with epoxy and place its corresponding sliced piece snugly around its outside. Hold the edges of the sliced pieces together with tape until the epoxy cures. Epoxy the launch lug/balsa strip cross assembly inside the BT-50 section.

28. Choose one side of the sighting arm to be the top and draw a line lengthwise along its center. Glue a strip of 3/16″ balsa flat in the middle of this line at the handle (rear) end of the sighting arm. Epoxy the rear (BT-50) sight centered on top.

29. Epoxy the front (crosshair) sight on the sighting arm centerline at the front end of the arm.

30. Cut a 4 1/4″ length of 15/32″ brass tubing and deburr the outside of each end.

31. Cut a 1 1/2″ square of 1/16″ rubber gasket material, then cut a 1/2″ square out of its center.

32. If improved weatherproofing is desired, coat all wood pieces, especially the Masonite and its edges, with an exterior wood sealer/primer.

33. Assemble the various pieces of the tracker as indicated in the drawings.

34. Cut two 3″ lengths of 0.047″ music wire for use as angle pointers. Drill a 3/64″ hole near the bottom edge of the rear side of the top redwood block. Bend the wire as necessary to bring it close to the azimuth protractor scale. Epoxy it in position.

35. Drill another 3/64″ hole in the sighting arm on the side toward the elevation protractor and 1″ forward of the handle. Put a 90° bend in the second piece of music wire so it will be near the angle scale. Do *not* epoxy it in position in this hole, however.

36. Attach the tracker to a steady tripod. Adjust the tripod and the wingnut/spring assemblies on the tilting table part of the tracker until the upper Masonite plate is exactly horizontal. Use a bubble level to determine this.

37. Put a bubble level at one end of the sighting arm and tilt the arm up or down as necessary until it is as close to horizontal as possible. Rotate it slowly in azimuth to ensure that the level continues to indicate it remains horizontal. Tighten the wing nut to lock the sighting arm temporarily in this position.

38. Epoxy the elevation pointer in a position where it reads 0°. It may be necessary to support it in this position with a small epoxied-on wood block. *This pointer must not be able to move.*

39. If you wish to do so, remove the sighting arm and spray paint it flat black to improve the sight contrast and reduce glare.

40. The Triple-Track Tracker is now ready for years of use.

Appendix VI

TWO-STATION ALT-AZIMUTH TRACKING DATA REDUCTION PROGRAM MRDR-2

NOTE: The following program is written in BASIC. The particular version of BASIC for your computer may differ slightly in command structure; check your computer's instruction manual.

```
10 REM MRDR-2.BAS by G. Harry Stine
20 REM Model Rocket Tracking Data Reduction Program
30 REM Program Copyright (C) 1993 by G. Harry Stine
40 REM May be freely copied and used but not sold
50 REM Data Input
60 DEFSNG A-Z
70 PRINT "Input baseline length in meters:";B
80 INPUT B
90 PRINT "Input Tracking East Azimuth angle:";A1
100 INPUT A1
110 PRINT "Input Tracking East Elevation angle:";E1
120 INPUT E1
130 PRINT "Input Tracking West Azimuth angle:";A2
140 INPUT A2
150 PRINT "Input Tracking West Elevation angle:";E2
160 INPUT E2
170 A3 = A1 + A2
180 R = .0174533
190 A1 = A1 *R
200 A2 = A2 *R
210 E1 = E1 *R
220 E2 = E2 *R
230 A3 = A3 *R
240 REM Average Altitude Calculations
250 V = B/SIN(A3)
260 X1 = V *TAN(E1) *SIN(A2)
270 X2 = V *TAN(E2) *SIN(A1)
280 PRINT "Altitude #1 = "; X1; " meters"
290 PRINT "Altitude #2 = "; X2; " meters"
300 X3 = (X1 + X2)/2
310 PRINT "Average Altitude = "; X3; " meters"
320 X4 = X3 *3.28084
330 PRINT "Average Altitude = "; X4; " feet"
340 REM 10% Closure Calculations
350 P1 = ((X3-X1)/X3) *100
360 IF P1 < 0 THEN P1 = P1 *(-1)
```

```
370 PRINT "Error: "; P1; "%"
380 IF P1 > 10 GOTO 410
390 PRINT "Track Closed"
400 GOTO 420
410 PRINT "No Closure"
420 END
```

Appendix VII

THREE-STATION ELEVATION-ANGLE-ONLY TRACKING DATA REDUCTION TABLES

Please refer to Figure 17-10 on page 281. These tables have been computed for three stations located in a straight line with the two end trackers located 100 meters (328.1 feet) from the middle tracker, giving a total baseline length of 200 meters (656.2 feet).

STEP 1

From Table 12:

End Tracker No. 1 elevation angle: _____
End Tracker No. 2 elevation angle: _____

Middle Tracker elevation angle: _____

End Tracker Number, Table 12: _____
End Tracker Number, Table 12: _____
Add Together: _____
Middle Tracker Number: _____
Subtract: _____

TABLE 12

Angle	End Tracker	Middle Tracker	Angle	End Tracker	Middle Tracker
1	3282.139	6564.279	26	4.203	8.407
2	820.034	1640.069	27	3.851	7.703
3	364.089	728.179	28	3.537	7.074
4	204.509	409.018	29	3.254	6.509
5	130.646	261.292	30	2.999	5.999
6	90.523	181.046	31	2.769	5.539
7	66.330	132.660	32	2.561	5.122
8	50.628	101.256	33	2.371	4.742
9	39.863	79.726	34	2.197	4.395
10	32.163	64.326	35	2.039	4.079
11	26.466	52.932	36	1.894	3.788
12	22.133	44.267	37	1.761	3.522
13	18.761	37.523	38	1.638	3.276
14	16.086	32.172	39	1.524	3.049
15	13.928	27.856	40	1.420	2.840
16	12.162	24.324	41	1.323	2.646
17	10.698	21.396	42	1.233	2.466

(continues)

TABLE 12
(continued)

Angle	End Tracker	Middle Tracker	Angle	End Tracker	Middle Tracker
18	9.472	18.944	43	1.149	2.299
19	8.434	16.868	44	1.072	2.144
20	7.548	15.097	45	0.999	1.999
21	6.786	13.572	46	0.932	1.865
22	6.126	12.252	47	0.869	1.739
23	5.550	11.100	48	0.810	1.621
24	5.044	10.089	49	0.755	1.511
25	4.598	9.197	50	0.704	1.408
51	0.655	1.311	71	0.118	0.237
52	0.610	1.220	72	0.105	0.211
53	0.567	1.135	73	0.093	0.186
54	0.527	1.055	74	0.082	0.164
55	0.490	0.980	75	0.071	0.143
56	0.454	0.909	76	0.062	0.124
57	0.421	0.843	77	0.053	0.106
58	0.390	0.780	78	0.045	0.090
59	0.361	0.722	79	0.037	0.075
60	0.333	0.666	80	0.031	0.062
61	0.307	0.614	81	0.025	0.050
62	0.282	0.565	82	0.019	0.039
63	0.259	0.519	83	0.015	0.030
64	0.237	0.475	84	0.011	0.022
65	0.217	0.434	85	0.007	0.015
66	0.198	0.396	86	0.004	0.009
67	0.180	0.360	87	0.002	0.005
68	0.163	0.326	88	0.001	0.002
69	0.147	0.294	89	0.000	0.000
70	0.132	0.264	90	0.000	0.000

STEP 2

From Table 13:

Look up the "Subtract" number obtained above in the "Sum of Values" column of Table 13. "Height" number opposite "Sum of Values" number is the achieved altitude in meters.

Sum of values: _____ height: _____ meters.

TABLE 13

Sum of Values	Height	Sum of Values	Height	Sum of Values	Height	Sum of Values	Height
20000.000	1	165.289	11	45.351	21	20.811	31
5000.000	2	138.888	12	41.322	22	19.531	32
2222.222	3	118.343	13	37.807	23	18.365	33
1250.000	4	102.040	14	34.722	24	17.301	34
800.000	5	88.888	15	32.000	25	16.326	35
555.555	6	78.125	16	29.585	26	15.432	36
408.163	7	69.204	17	27.434	27	14.609	37
312.500	8	61.728	18	25.510	28	13.850	38
246.913	9	55.401	19	23.781	29	13.149	39
200.000	10	50.000	20	22.222	30	12.500	40

TABLE 13

Sum of Values	Height	Sum of Values	Height	Sum of Values	Height	Sum of Values	Height	Sum of Values	Height
11.897	41	2.415	91	1.005	141	0.548	191	0.344	241
11.337	42	2.362	92	0.991	142	0.542	192	0.341	242
10.816	43	2.312	93	0.978	143	0.536	193	0.338	243
10.330	44	2.263	94	0.964	144	0.531	194	0.335	244
9.876	45	2.216	95	0.951	145	0.525	195	0.333	245
9.451	46	2.170	96	0.938	146	0.520	196	0.330	246
9.053	47	2.125	97	0.925	147	0.515	197	0.327	247
8.680	48	2.082	98	0.913	148	0.510	198	0.325	248
8.329	49	2.040	99	0.900	149	0.505	199	0.322	249
8.000	50	2.000	100	0.888	150	0.500	200	0.320	250
7.689	51	1.960	101	0.877	151	0.495	201	0.317	251
7.396	52	1.922	102	0.865	152	0.490	202	0.314	252
7.119	53	1.885	103	0.854	153	0.485	203	0.312	253
6.858	54	1.849	104	0.843	154	0.480	204	0.310	254
6.611	55	1.814	105	0.832	155	0.475	205	0.307	255
6.377	56	1.779	106	0.821	156	0.471	206	0.305	256
6.155	57	1.746	107	0.811	157	0.466	207	0.302	257
5.945	58	1.714	108	0.801	158	0.462	208	0.300	258
5.745	59	1.683	109	0.791	159	0.457	209	0.298	259
5.555	60	1.652	110	0.781	160	0.453	210	0.295	260
5.374	61	1.623	111	0.771	161	0.449	211	0.293	261
5.202	62	1.594	112	0.762	162	0.444	212	0.291	262
5.039	63	1.566	113	0.752	163	0.440	213	0.289	263
4.882	64	1.538	114	0.743	164	0.436	214	0.286	264
4.733	65	1.512	115	0.734	165	0.432	215	0.284	265
4.591	66	1.486	116	0.725	166	0.428	216	0.282	266
4.455	67	1.461	117	0.717	167	0.424	217	0.280	267
4.325	68	1.436	118	0.708	168	0.420	218	0.278	268
4.200	69	1.412	119	0.700	169	0.417	219	0.276	269
4.081	70	1.388	120	0.692	170	0.413	220	0.274	270
3.967	71	1.366	121	0.683	171	0.409	221	0.272	271
3.858	72	1.343	122	0.676	172	0.405	222	0.270	272
3.753	73	1.321	123	0.668	173	0.402	223	0.268	273
3.652	74	1.300	124	0.660	174	0.398	224	0.266	274
3.555	75	1.280	125	0.653	175	0.395	225	0.264	275
3.462	76	1.259	126	0.645	176	0.391	226	0.262	276
3.373	77	1.240	127	0.638	177	0.388	227	0.260	277
3.287	78	1.220	128	0.631	178	0.384	228	0.258	278
3.204	79	1.201	129	0.624	179	0.381	229	0.256	279
3.125	80	1.183	130	0.617	180	0.378	230	0.255	280
3.048	81	1.165	131	0.610	181	0.374	231	0.253	281
2.974	82	1.147	132	0.603	182	0.371	232	0.251	282
2.903	83	1.130	133	0.597	183	0.368	233	0.249	283
2.834	84	1.113	134	0.590	184	0.365	234	0.247	284
2.768	85	1.097	135	0.584	185	0.362	235	0.246	285
2.704	86	1.081	136	0.578	186	0.359	236	0.244	286
2.642	87	1.065	137	0.571	187	0.356	237	0.242	287
2.582	88	1.050	138	0.565	188	0.353	238	0.241	288
2.524	89	1.035	139	0.559	189	0.350	239	0.239	289
2.469	90	1.020	140	0.554	190	0.347	240	0.237	290

(continues)

TABLE 13
(continued)

Sum of Values	Height	Sum of Values	Height	Sum of Values	Height	Sum of Values	Height	Sum of Values	Height
0.236	291	0.171	341	0.130	391	0.102	441	0.082	491
0.234	292	0.170	342	0.130	392	0.102	442	0.082	492
0.232	293	0.169	343	0.129	393	0.101	443	0.082	493
0.231	294	0.169	344	0.128	394	0.101	444	0.081	494
0.229	295	0.168	345	0.128	395	0.100	445	0.081	495
0.228	296	0.167	346	0.127	396	0.100	446	0.081	496
0.226	297	0.166	347	0.126	397	0.100	447	0.080	497
0.225	298	0.165	348	0.126	398	0.099	448	0.080	498
0.223	299	0.164	349	0.125	399	0.099	449	0.080	499
0.222	300	0.163	350	0.125	400	0.098	450	0.080	500
0.220	301	0.162	351	0.124	401	0.098	451	0.079	501
0.219	302	0.161	352	0.123	402	0.097	452	0.079	502
0.217	303	0.160	353	0.123	403	0.097	453	0.079	503
0.216	304	0.159	354	0.122	404	0.097	454	0.078	504
0.214	305	0.158	355	0.121	405	0.096	455	0.078	505
0.213	306	0.157	356	0.121	406	0.096	456	0.078	506
0.212	307	0.156	357	0.120	407	0.095	457	0.077	507
0.210	308	0.156	358	0.120	408	0.095	458	0.077	508
0.209	209	0.155	359	0.119	409	0.094	459	0.077	509
0.208	310	0.154	360	0.118	410	0.094	460	0.076	510
0.206	311	0.153	361	0.118	411	0.094	461	0.076	511
0.205	312	0.152	362	0.117	412	0.093	462	0.076	512
0.204	313	0.151	363	0.117	413	0.093	463	0.075	513
0.202	314	0.150	364	0.116	414	0.092	464	0.075	514
0.201	315	0.150	365	0.116	415	0.092	465	0.075	515
0.200	316	0.149	366	0.115	416	0.092	466	0.075	516
0.199	317	0.148	367	0.115	417	0.091	467	0.074	517
0.197	318	0.147	368	0.114	418	0.091	468	0.074	518
0.196	319	0.146	369	0.113	419	0.090	469	0.074	519
0.195	320	0.146	370	0.113	420	0.090	470	0.073	520
0.194	321	0.145	371	0.112	421	0.090	471	0.073	521
0.192	322	0.144	372	0.112	422	0.089	472	0.073	522
0.191	323	0.143	373	0.111	423	0.089	473	0.073	523
0.190	324	0.142	374	0.111	424	0.089	474	0.072	524
0.189	325	0.142	375	0.110	425	0.088	475	0.072	525
0.188	326	0.141	376	0.110	426	0.088	476		
0.187	327	0.140	377	0.109	427	0.087	477		
0.185	328	0.139	378	0.109	428	0.087	478		
0.184	329	0.139	379	0.108	429	0.087	479		
0.183	330	0.138	380	0.108	430	0.086	480		
0.182	331	0.137	381	0.107	431	0.086	481		
0.181	332	0.137	382	0.107	432	0.086	482		
0.180	333	0.136	383	0.106	433	0.085	483		
0.179	334	0.135	384	0.106	434	0.085	484		
0.178	335	0.134	385	0.105	435	0.085	485		
0.177	336	0.134	386	0.105	436	0.084	486		
0.176	337	0.133	387	0.104	437	0.084	487		
0.175	338	0.132	388	0.104	438	0.083	488		
1.174	339	0.132	389	0.103	439	0.083	489		
0.173	340	0.131	390	0.103	440	0.083	490		

SAMPLE NAR SECTION BYLAWS

These are sample bylaws. Individual sections may wish to alter them or add to them, due to local circumstances. Please *do not* merely fill in the blanks of this sheet and forward it to NAR Headquarters for approval; make your own copies, and be sure you have enough for your members and a copy for NAR Headquarters. The purpose of these sample bylaws is to provide a guide for each group in drawing up its own bylaws. All bylaws and amendments thereto must be approved in writing by NAR Headquarters.

Article 1, Name: The name of this organization shall be the _____
_____
Section of the National Association of Rocketry.

Article 2, Purpose: It shall be the purpose of this Section to (a) aid and abet the aims and purposes of the NAR in (locale) _____
_____,
(b) to operate and maintain a model rocket range in accordance with the NAR Standards and Regulations, (c) to hold meetings for the purpose of aiding and encouraging all those interested in rocketry, and (d) to engage in other scientific, educational, or related activities as the NAR, the Section, or the Section Board of Directors may from time to time deem necessary or desirable in connection with the foregoing.

Article 3, Membership: All members of this Section shall be NAR members in good standing who reside in _____
(locale) _____

Article 4, Dues: Dues shall be $_____
per year, payable in advance. These Section dues are separate and distinct from national dues paid to the NAR. All dues monies shall be kept in a Gen-

eral Fund by the Secretary-Treasurer and shall be paid out by him only on order of the Section Board of Directors. Special assessments may be levied by a majority vote of the members present and voting at any meeting of the Section, provided notice of such intent is given in writing to each member at least five days preceding such a meeting.

Article 5, Meetings: Meetings of the Section shall be held at least _____ times per year at times and places designated by the Section Board of Directors. Operation of the rocket range shall not be considered a meeting. A quorum shall consist of 50% of the membership of the Section. Meetings shall be conducted and governed by *Robert's Rules of Order, Revised.*

Article 6, Board of Directors: The Board of Directors of this Section shall consist of the three officers, one member at large, and a Senior Member of the NAR, who shall be designated by the NAR as Section Advisor.

Article 7, Officers: The officers of this Section shall consist of a President, a Vice-President, and a Secretary-Treasurer, all of whom shall be members of the Section and of the NAR.

Article 8, Elections: Elections of officers and members of the Board of Directors shall take place at the first meeting of the calendar year. All officers and members of the Board shall serve a term of one year. Vacancies in offices and on the Board shall be filled by nomination and election of a Section member to fill the unexpired term of office and shall take place at the Section meeting at which the vacancy is announced. Nominations for all elections shall be made from the floor, and the candidate having the largest number of votes shall be elected.

Article 9, Committees: There shall be three Standing Committees of the Section, plus such additional committees as the Board of Directors may from time to time deem necessary or desirable. The Standing Committees are as follows:

(a) Operations Committee shall be in charge of the Section's model rocket range, shall monitor the experimental technical activities of the Section members, and shall act as safety inspectors. The Chairman of this Committee shall be a Senior Member of the NAR in good standing and shall act as Range Safety and Control Officer under the NAR Official Standards and Regulations.

(b) The Contests and Records Committee shall be in charge of all arrangements for contests and shall monitor all national-record attempts by Section members. The Committee shall contain at least one Leader Member of the NAR.

(c) The Activities Committee shall be in charge of making all arrangements for all Section meetings, for conducting membership campaigns, and for carrying on public relations.

(d) The Section President shall be an ex-officio member of all committees.

Article 10, Amendments: These bylaws may be amended by a two-thirds vote of those Section members present and voting at any meeting of the Section, provided written notice of the pending amendment has been sent to the membership of the Section at least five days in advance of such meeting. No amendment of these bylaws shall be in force until approved by NAR Headquarters.

Adopted: _____

Approved by NAR Headquarters: _____

INDEX

ABOUT THE AUTHORS

G. Harry Stine (1928–1997) is probably the person most responsible for starting model rocketry and guiding it through its early years as it grew into a highly respected worldwide aerospace hobby and sport. In 1957, he founded the National Association of Rocketry (NAR) and started the first model rocket company, Model Missiles, Inc. An engineer, science writer, and science fiction author, he wrote over sixty books. His many science fiction stories once appeared under the pen name Lee Correy, until he began writing fiction under his own name in 1985. He has always used his given name for all his nonfiction.

This book, *Handbook of Model Rocketry*, has been cited as one of his most influential works and is in its seventh edition as the bible and official reference book of the entire hobby. The success of this book has prompted many other books on model rocketry here and abroad, but it remains the updated original and most used book about model rocketry.

Stine not only founded the NAR but served for ten years as its president and was for many years an honorary member and an honorary trustee of the NAR. For eleven years he served as the president of the Space Model Subcommittee of the Federation Aeronautique Internationale (FAI) in Paris. He was also a member and chairman of the Technical Committee on Pyrotechnics of the National Fire Protection Association. He was the U.S. senior national champion model rocketeer four times and guided his two daughters and son to the junior championships. He held the NAR membership and sporting license number NAR #2.

His list of honors and memberships is long: fellow of the Explorers Club, fellow of the British Interplanetary Society, associate fellow of the American Institute of Aeronautics and Astronautics, scientific leader member of the Academy of Model Aeronautics, and member of the New York Academy of Sciences, among others. He is listed in *Who's Who in America*. In 1969, he received a special award from the Hobby Industry Association for his role in establishing the model rocket industry and a silver medal as one of fifty U.S. space pioneers honored by the U.S. Army Association.

Thousands of model rocketeers knew him as "The Old Rocketeer."

Bill Stine, NAR#24, is G. Harry's son. Bill has been active in all aspects of the hobby of model rocketry since early childhood and has continued to carry forward the legacy left to him by his father. Bill is the founder and president of Quest Aerospace, Inc. Founded in 1991, Quest quickly became one of the two leading model rocket companies. Prior to founding Quest, Bill worked for Model Rectifier Corporation (MRC) and was responsible for the complete design and development of the Concept II line of model rocket products. Prior to his involvement with MRC, he was vice president of marketing and operations for ENERTEK, Inc. (now Aerotech, Inc.), a pioneer manufacturer of large model rocket products in Phoenix, Arizona. He worked in the research and development and marketing departments of Centuri Engineering and Estes Industries from 1978 to 1981. He has designed and built rocket models professionally for two Hollywood movies, General Dynamics Space Systems, and Motorola's Iridium. He has served as an industry expert committee member of the National Fire Protection Association Committee on Pyrotechnics since 1989. He has been a reserve national champion competitor with the National Association of Rocketry. In 1999, he founded the Model Rocket Museum to further document and preserve the history of the hobby. Today, Bill travels the world giving model rocket workshops for teachers, flying model rockets, and as taught by his father, "paying forward."

He lives in Cave Creek, Arizona, with his wife, Lisa (whom he met in 1977 at a model rocket competition), their two daughters, and two golden retrievers (rocket recovery crew).